George William Balfour

The Senile Heart

It's Symptoms and Treatment

George William Balfour

The Senile Heart
It's Symptoms and Treatment

ISBN/EAN: 9783744675994

Printed in Europe, USA, Canada, Australia, Japan

Cover: Foto ©berggeist007 / pixelio.de

More available books at **www.hansebooks.com**

THE SENILE HEART

Its Symptoms, Sequelæ, and Treatment

BY

GEORGE WILLIAM BALFOUR, M.D. (St. And.)
LL.D. (Ed.), F.R.C.P.E., F.R.S.E.

Consulting Physician to the Royal Infirmary, to the Royal Hospital
for Sick Children, and to the Royal Public Dispensary,
Edinburgh; Consulting Physician to Leith Hospital,
formerly Physician to Chalmers Hospital,
Edinburgh, etc.; Member of the University Court of St. Andrews

Nascentes morimur, finisque ab origine pendet
—Manilius, *Astronomicon*, iv. 16

New York
THE MACMILLAN COMPANY
LONDON: MACMILLAN & CO., Ltd.
1898

All rights reserved

COPYRIGHT, 1894,
BY MACMILLAN AND CO.

Set up and electrotyped August, 1894.   Reprinted February, 1898.

Norwood Press
J. S. Cushing & Co. — Berwick & Smith
Norwood Mass. U.S.A.

PREFACE

DISEASE often deranges the mechanism of the cardiac valves, and thus places an actual or constructive obstacle in the way of the onward flow of the blood. To maintain the circulation under these conditions the myocardium must hypertrophy — the heart necessarily enlarges. This we all know. But few realize that the loss of elasticity, and other changes which the arterial system undergoes, during our progress from youth to age, also cause a hindrance to the onward flow of the blood which has to be compensated in a similar manner. In late life, and without any history of previous disease, the heart is often found to be enlarged, and this enlargement is under these circumstances said to be idiopathic. But enlargements of the heart form no exceptions to the universal law that there is no effect without an antecedent cause.

Owing to the changes in the vascular system

just referred to, no heart reaches advanced age without some degree of enlargement. This trifling enlargement is of slow growth, gives rise to no symptoms, and is only found when looked for. But when after middle life distressing symptoms attract attention to the heart, in by far the larger proportion of cases there is discoverable no history of any antecedent myocarditis or other disease, but the symptoms are entirely due to disturbance of the nutrition or of the innervation of the myocardium interfering with and modifying the normal senile enlargement of the heart.

By far the most widespread and most interesting varieties of cardiac disease are to be found in this connection, while the comfort and longevity of many depend upon a clear understanding of the various causes which contribute to such modifications of the SENILE HEART, and an appropriate treatment of the many distressing symptoms so often associated with it.

17 WALKER STREET, EDINBURGH,
    February, 1894.

CONTENTS

CHAPTER		PAGE
I.	INTRODUCTORY	1
II.	HOW THE HEART IS AFFECTED BY AGE	21
III.	SYMPTOMS AND SIGNS OF THE SENILE HEART	35
IV.	PALPITATION, TREMOR CORDIS, TACHYCARDIA	63
V.	BRADYCARDIA, AND DELIRIUM CORDIS	91
VI.	ANGINA PECTORIS	115
VII.	CONCOMITANTS AND SEQUELÆ OF THE SENILE HEART. GOUT	157
VIII.	CONCOMITANTS AND SEQUELÆ OF THE SENILE HEART. GLYCOSURIA, GOUTY KIDNEYS	187
IX.	THE THERAPEUTICS OF THE SENILE HEART. GENERALITIES	214
X.	THE THERAPEUTICS OF THE SENILE HEART. EXERCISE AND DIET	231
XI.	THE THERAPEUTICS OF THE SENILE HEART. DRUGS LIKELY TO BE USEFUL, AND HOW TO USE THEM	258
XII.	THE PROGNOSIS OF SPECIAL SYMPTOMS. RECAPITULATION OF TREATMENT WITH SPECIAL REFERENCE TO SYMPTOMS	286

ILLUSTRATIONS

Figure		Page
1.	Innervation of Heart	38
2.	Sphygmogram of Feeble and Irregular Pulse	46
3. 4.	Sphygmogram of Irregular Pulses in Dilated Hearts	49
5.	Sphygmogram of Tachycardiac Pulse	78
6.	Sphygmogram of Hemisystolic Bradycardia	106
7.	Sphygmogram of True Bradycardia	107
8.	Ridged Gouty Nail	177
9.	Furrowed Nail	177
10.	Heberden's Knobs	179
11.	Haygarth's Nodosities	181

THE SENILE HEART

CHAPTER I

INTRODUCTORY

THE late Sir Robert Christison, in his first report on the emerged risks of the Standard Assurance Company, stated that the statistics he was dealing with seemed to show that but few even of the aged die from natural decay, but "mostly from some specific disease, just like younger persons."[1] And he also said that the term "gradual decay," when used as an explanation of the cause of death of even old people, was "little else than an admission of ignorance."[2] It is consolatory to have such high authority for believing that at the most advanced ages death is not due to age alone, but to disease; because we always personify disease, we feel that we may escape it, we can fight it,

Death rarely due to age alone.

[1] *Monthly Journal of Medical Science*, August, 1853, p. 109.
[2] *Op. cit.*, p. 110.

and often overcome it; but age is the "carle dour" to whom we must all succumb. Hence a well-founded belief that disease and not age has been the cause of death, even at the most advanced ages hitherto recorded, is fraught with the hope that science, if not luck, may make the patriarchal ages again our own, and that the mantle of Methuselah may yet fall upon the shoulders of his nineteenth century successors. From Sir Robert's point of view there is no reason to regard this as impossible; it even looms in the future as vaguely probable. But there is another point from which the lookout is not quite so hopeful.

Yet as man's life is limited, age must have an important effect.

Many years ago an old writer recorded the traditional experience of his time in these memorable words: "The days of our years are threescore years and ten; and if by reason of strength they be fourscore years, yet is their strength labour and sorrow; for it is soon cut off, and we fly away."[1] The Psalmist does not set up threescore and ten as an age to which all must attain and which none may exceed, but merely states it as the average limit of a full and complete life beyond which but few may pass, and that only "by reason of strength," which soon withereth away.

[1] Psalm xc., verse 10.

And to-day the same story is repeated in the prosaic pages of the Registrar-General, with all the emphatic truthfulness of nineteenth century statistics. "Of 100,000 born in this country, it has been ascertained that one-fourth die before they reach their fifth year; and one-half before they have reached their fiftieth year. Eleven hundred will reach their ninetieth. And only two persons out of the 100,000 — like the last barks of an innumerable convoy — will reach the advanced and helpless age of one hundred and five."[1] Of the many millions born, only isolated exceptions attain great ages, and all are ultimately entombed in the urns and sepulchres of mortality, for "time like an overwhelming flood bears all his sons away." Tradition and statistics are thus agreed that few of those born attain the age of threescore and ten, and that beyond that age they die off so rapidly that seventy years may be practically regarded as the extreme limit of even a long life.

The life of the human body having thus an end as well as a beginning, it may be identified

---

[1] The above quotation will be found at p. 24 of Smiles' work on *Thrift*. London, John Murray, 1886. It is based on the Life-tables of the Registrar-General; one of the latest of these makes one-half die before forty-seven. *Vide* Table B., p. vii., *Supplement to Annual Report*, 1885.

with development, and all its phases inseparably linked with structural change.[1]

Life is linked with development, and consequently terminates necessarily and naturally in death.

We are so much accustomed to associate development with growth merely, that it seems a somewhat startling proposition to connect it also with decay, and but few of us are sufficiently educated to regard the development of the body as naturally ending only with its death.[2] But this conception of the nature of development involves also the idea that as there are developmental phenomena initial in character, so there must also be similar phenomena which are terminal. And this brings with it matter for serious consideration; for with it comes also the reflection that terminal phenomena may not always be restricted to that advanced age, to which alone they seem to be appropriate. For just as we may have a precocious development,[3] so we may also

---

[1] "Jede Function ist an mechanische Veränderungen der Substanz geknüpft."—Virchow, *Vier Reden über Leben und Kranksein*, Berlin, 1863, p. 96.

[2] "Development and Life are, strictly speaking, one thing; though we are accustomed to limit the former to the progressive half of life, and to speak of the retrogressive half as decay, considering an imaginary resting-point between the two as the adult or perfect state."—Huxley, *British and Foreign Medical Review*, October, 1853, p. 305.

[3] In the *British Medical Journal* for 6th February, 1886, p. 263, mention is made of a child three and a half years old

have a premature decay. The development of man, in the only true sense of the expression, is a physiological process, dependent upon tissue change and not on years, and it may attain its natural termination before, as well as after, the conventional threescore and ten. To employ the expression "gradual decay" as an indication that death has occurred from age alone may often

who looked like a boy of ten or twelve years of age, and in whom puberty commenced at the early age of eighteen months. Many similar cases have been recorded by various authors from Seneca and Pliny downwards. Prematurity of development and precocity of growth are essentially distinct, though there is a close bond of union between them. Thus females not infrequently menstruate prematurely, but precocity of growth in that sex is so rare that Geoffroy Saint-Hilaire has only recorded two cases of combined precocity of growth and prematurity of development among females. Among males, on the other hand, many such cases have been recorded, and in them premature development of the genital organs is almost invariably associated with precocity of growth. Some of the cases recorded have been very remarkable. Sauvages, *Hist. de l'Acad. de* 1666 à 1669, t. ii., p. 43, has given full particulars of a boy of six who was five feet high and broad in proportion. His growth was so rapid that it could almost be seen; he had a beard, looked like a man of thirty, and had every indication of perfect puberty. He had a full, deep bass voice, and his extraordinary strength fitted him for all country work. At five he could carry any distance three measures of rye weighing 84 pounds; and at six years and a few months he could easily carry on his shoulders burdens weighing 150 pounds. But he did not become a giant as everybody expected; he soon got feeble, deformed, and almost an idiot. *Vide Histoire des Anomalies*, par M. Isidor Geoffroy Saint-Hilaire, Paris, 1832, Vol. i., p. 197, etc. *Vide* also note 1, p. 12.

enough be little else than "an admission of ignorance," but to confound terminal phenomena with disease is worse; it is the unconscious revelation of an ignorance which is not admitted.

The superficial observer, who fixes his attention upon gradual decay alone, sees but little of death from age; but he who recognizes the existence, the nature, and the importance of terminal phenomena, not only sees many deaths from age, but is often privileged to ward off for long the ultimate and inevitable end. The linking of growth with decay as part of the development of man opens up striking views as to the gradual evolution of terminal phenomena, as well as of the importance of the early recognition of their first beginnings, and of the various modes in which they threaten life. Because in this matter perturbative medicine and heroic measures can do no good, and may do much harm. To be of any use at all, we must put ourselves in nature's place and work as nature works.[1] Here or there we

Obsta principiis an important aid in prolonging life.

[1] Our lives are but a bundle of consequences; our present is but the outcome of our past. It is by trifling advantages, momentarily minute and imperceptible, that nature either worsens or improves the status of our vitality, and it is by securing these trifling advantages and turning them to the good of our patient that vital declension is averted, and chronic ailments remedied when that is possible. *Vide* Darwin, *Origin of Species*, and Balfour's *Introduction to the Study of Medicine*, A. & C. Black, 1865, p. 237.

discover some trifling failure, which, like the "little rift within the lute," threatens serious disaster erelong; but appropriate remedies, timely applied, and long persevered with, may enable us to avert this disaster, and by the slow accumulation of petty advantages change the commencement of decay into the renewal of strength. In this way we shall more certainly prolong our life, and secure comfort in existence, than by the unguents, the hot baths, and the elixir vitæ, by which our forefathers sought to emolliate the rigidity of age, and to add a fresh stock of vital force to that which was fast wearing away.[1]

We are born with potentialities, not powers; we have no store of energy upon which to draw, which may be wasted, and which must daily diminish. It is well for us that it is so, as we are all so apt to squan- *Life a form of energy.*

---

[1] The notion that desiccation is the cause of age, in the obnoxious sense of the word, was widely prevalent in early times. It gave rise to the story of the rejuvenation of Pelias in Medea's caldron, to which Lord Bacon refers as an instance of the utility of warm bathing in warding off age. Desiccation as a cause of age is also referred to by Galen in his treatise *De Sanitate Tuendo* and *De Marasmo;* and Haller actually states that fishes live long because their bones are soft and cartilaginous. *Primæ Lineæ*, § 972. Early indications of all the most modern ideas of preserving vitality by drinking hot fluids, clothing in woollen garments, and feeding on peptonized aliments, are to be found in Lord Bacon's *Historia Vitæ et Mortis*, Spalding's edition, London, 1858, Vols. ii. and v.

der our energy in work or play, or to have it wasted for us by disease, that had we only a fixed amount on which to draw, but few of us would live to old age, and not so many as now even to middle life. As it is, all our energy comes from without; we take it in as food in the Potential form, and we transform it into the Kinetic form by means of the oxygen circulating in our blood. Every act of life, every pulse that beats within us, every thought we think, involves this transformation of energy, even though no apparent movement indicates the presence of voluntary life. That we may continue to live, the products of this chemical action must be removed, and the used-up waste replaced by fresh oxidizable material; and our organism is so constructed that for a time this goes on continuously. The latest definition of life is based upon these facts, for "the continuous adjustment of internal relations to external relations"[1] is obviously but a concise statement of the conditions necessary for the continuous manifestation of living action, and not a definition of life itself. Though it therefore fails the metaphysician, this definition is practically sufficient for the physician, who deals only with the physics involved. And

Life defined.

---

[1] *The Principles of Biology*, by Herbert Spencer. Williams and Norgate, London, 1865, Vol. i., p. 80.

we may also accept the converse, that death is the result of a "failure to balance ordinary external actions by ordinary internal actions."[1] *Death defined.* But food and oxygen remaining plentiful, as we may assume they ordinarily do, there seems no reason why assimilation, oxidation, and the genesis of force should not go on forever, or until some cataclysmic change in our environment should disturb the balance of internal and external actions. It is within ourselves, therefore, that we must seek for that change which causes these processes of assimilation, oxidation, and the genesis of force, gradually to fall out of correspondence with the relations between oxygen and food, and the absorption of heat by the environment, which happens in old age, and is the cause of death by natural decay.[2]

In days gone by the hypothesis of a gradual decrease of vital force was supposed to explain the enigma of gradual decay.[3] But vital force is but another name for the sum of all the vital actions of the frame; *Old idea of vital force.* and to point out that these are decreasing is,

---

[1] Spencer, *op. cit.*, p. 89.
[2] Spencer, *op cit.*, p. 88.
[3] "La gene de l'influence vital s'accroit sans cesse." — Cabanis. "That considerable differences exist in the stock of vitality originally imparted to the frame in different individuals cannot be doubted, some being destined to a shorter, and others to a

indeed, to indicate that the organism is dying, but is no explanation of why it dies. Failure in the genesis of force is only an indication of failure in oxidation or assimilation.

So, too, impoverishment of the blood, which has been regarded as the cause of the gradual failure in the aged,[1] is itself due to imperfect assimilation, and leads to imperfect oxidation and consequent failure in the genesis of force. Imperfect assimilation is, doubtless, one of the most important links in the chain of causes which lead to the general decay of the bodily frame. The difficulty is to say where this chain begins; for all the functions of the body are so linked together that there is not one of them which can be called primary, upon which, when it fails, may be laid the blame of initiating the decay of all the others.[2]

Causes of failure in the genesis of force.

The more, indeed, we investigate the phenomena of decay, the more clearly do we see that this does not arise from any failure of the sources of potential energy, but solely from the inability

---

longer, term of existence." — Roget, article "Age," in the *Cyclopedia of Practical Medicine*, etc.

[1] *Vide* articles on "Age," in the *Cyclopedia of Anatomy and Physiology*, by Symonds, p. 82; and in the *Cyclopedia of Practical Medicine*, by Roget, p. 39.

[2] Roget, *loc. cit.*; and Whytt, *On Vital Motions*, Edinburgh, 1751, p. 270.

of the organism to make use of those presented to it, because it has itself become effete as the direct and necessary result of development.

In early life the body grows through the abundance of the fluid food, with which every part is flushed. The characteristics of the systemic circulation upon which this flushing depends are, that the amplitude (calibre) of the large arteries is great in comparison to the size of the heart, and also to the length of the body. Hence there is a low blood pressure and a rapid pulse-rate. The large amount of fluid in the tissues, the abundant supply of nutriment, and the low blood pressure, coupled with the shorter time in which the whole circuit of the vascular system is traversed,[1] all favour the diffusion of the blood-plasma and the rapid growth of the body. These conditions prevail during early life, but are most marked during the first year. During early life the whole body grows in every part, but the growth of the arteries in calibre does not keep pace either with the growth of the body in length, or with the growth of the heart in amplitude and strength. The natural result is a gradual rise in the blood pressure, and an equally gradual slowing of the pulse-rate

Characteristics of the circulation in early life.

---

[1] Twelve seconds as against twenty-two in the adult. *Vide* Foster's *Physiology*, 1883, p. 685.

as growth and age increase, until in early manhood growth is completed, and the blood pressure reaches its highest norm. At this period the whole organism is full of life and vigour, and is at its best in respect of its capacity for bodily and mental exertion.[1] But as development progresses,

---

[1] *Vide Die Altersdisposition*, by Dr. F. W. Beneke, Marburg, 1879, pp. 7, 12, 14, and 18. About this time two events, the access of puberty and the cessation of growth, have a most important influence in the story of development. Beneke refers the access of puberty to the cessation of growth. The capillary system, during the second stage of life (7–15), ceases to grow as heretofore; the brain and large glands have attained nearly their full development, and the blood pressure, still rising in the whole arterial and capillary system, finds its outlet in the development of the sexual organs, the glands of the skin, and the growth of the hair always associated with maturity (*op. cit.*, p. 14). This explanation of the physics of development seems adequate enough so far as it goes. The great difficulty is to account for the cessation of growth at all in any part of the body. Geoffroy Saint-Hilaire (*op. cit.*, vol. i.) has some very pertinent and interesting remarks on this subject. He points out that dwarfs are imperfect individuals, usually impotent; their growth and development have both been arrested. But inasmuch as puberty alone puts an end to the growth of man, a dwarf may recommence his growth at any time, even up to old age; and Saint-Hilaire states that he himself had observed several instances of this (p. 190, note). Giants, on the other hand, are those in whom growth has continued because their sexual organs have been slowly and incompletely developed. Giants are mostly impotent, always feeble, feminine in aspect, and usually very shortlived (p. 192). In precocious children, the rapid growth being early directed to the sexual organs, the individual ceases to grow; from a gigantic child he may become, if he lives, a man of but moderate bulk (p. 192). So it may

the arterial coats slowly undergo a change of structure, by which they lose their elasticity, and become gradually converted into more or less approximately rigid tubes.[1] The effect of this loss

Changes in the vascular system through age.

happen that a child with very active nutrition may become either an imperfectly developed giant, or an early developed (precocious) youth of moderate bulk (pp. 193, 194). Saint-Hilaire merely states these as facts, without in any way attempting to account for them. He points out that there are various races of men famous for their bulk and stature, and others remarkable for their diminutive size; and that though food and other accidents of environment have an acknowledged influence in promoting and hindering growth, yet those races of varying size unquestionably owe more to heredity (however acquired) than they do to abundance of food, and a comfortable, easy life, or the reverse (pp. 240, 241). It is the physics alone of growth that concern us : could we know these perfectly, it would suffice ; meanwhile as these physics are closely involved with the progress and cessation of growth, it would be of great importance to discover why we ever cease to grow. Herbert Spencer says that growth is arrested " because the excess of absorbed over expended nutriment must, other things equal, become less as the size of the animal become greater." — *Principles of Biology*, Vol. i., p. 122. If this were the true reason, we should all be more nearly alike in size than we are. Moreover, though this seems a good enough reason for an animal ceasing to live when it attains a certain bulk, and thus seems applicable to giants, and explanatory of their short lives, it does not seem to be a sufficient reason why we should ever cease growing before we reach that extreme bulk, nor does it give any explanation of the very peculiar relations subsisting between growth and sexual development.

[1] " One common feature of old age is the conversion by such a change " — that is, by the replacement of a structured matrix

of resilience in the arterial coats is, that while these coats yield as formerly to the advancing blood-wave, they yield more slowly, and they do not recover themselves, so that the lumen of the arteries undergoes a gradual dilatation. The heart at the same time tends to fail, senile atrophy begins, and in the midst of our fullest life Death himself lays his finger upon all our organs. From the dilatation of the arteries there is a tendency to lowering of the blood pressure; and to this failure of the blood pressure has been ascribed that obsolescence of the capillaries which is the cause of the dry and wrinkled skin, the gray hair, and the cessation of the sexual functions, and which is so evident on the anatomical investigation of the organs themselves.[1] The result of this withering of the capillaries is, according to Beneke, by diminishing their area, to increase the peripheral resistance to the onward flow of the blood, and thus to raise the blood pressure within the arteries themselves, so that, notwithstanding dilatation of these vessels, the blood pressure in age is always greater than it is in early youth. This view of Beneke's is not, however, quite consistent with the physical facts;

---

by amorphous material — "of the supple, elastic arteries into rigid tubes." — Foster, *op. cit.*, p. 690. *Vide* also references on p. 20.   [1] Beneke, *op. cit.*, p. 24.

circumstances being alike, the increase in the arterial capacity would undoubtedly lower the blood pressure within them; but the circumstances of age are by no means those of youth. In youth the relatively large calibre of the arteries has no ill effect on the circulation, because though the blood pressure is not great, it is perfectly sufficient to keep up a steady and continuous flow into the capillaries. In age, however, the case is different; the loss of arterial elasticity, while it throws a greater strain upon the heart itself, makes the outflow into the capillaries approximately intermittent, and thus lowers the blood pressure within the capillary area, though it still remains high within the arteries themselves.[1] The cessation of active growth makes a large network of capillaries unnecessary, and the fall of the blood pressure within these vessels permits many of them to obsolesce. The result, therefore, is identical, though the steps by which it is reached are not exactly as Beneke has put them. By the time the heart has succeeded in permanently dilating the inelastic arteries, and has restored them to their former relative magnitude, the increase of the peripheral resistance, due to the withering of the capillaries, is sufficient to prevent any material lowering of the blood pressure from this cause. By and by,

[1] Foster, *op. cit.*, p. 132.

however, this is gradually brought about by weakening of the heart through failure of the genesis of force, due to failure of assimilation arising from withering of the capillaries in the skeletal muscles, as well as in all the glands of the body.

Decay is thus the necessary and final stage of development; and though it may not be possible to put a finger upon any special function or structure, and to say, Here decay commences, yet erelong we can positively say, *This* is the line along which decay is marching, and *here* is the structure in which we can earliest detect the withering effects of age.

From our earliest days the growth of our frame is accompanied by a gradual condensation of tissue, till the gelatinous pulp of the primitive embryo is converted into the withered old man. Every tissue partakes of this change: the skin becomes dry, flaccid, and wrinkled; the bones are denser and more brittle; the muscles participate in the condensation incident to the cellular tissue, which enters so largely into their composition; the muscular fibres themselves are more rigid, diminished in bulk, and impaired in contractility, so that they are less readily and less powerfully excited by stimuli.[1] Hence the shrunk shanks, tottering gait, and withered aspect of the aged

Tithonus a true type of protracted age.

---

[1] Roget, *op. cit.*, p. 40.

man which have crystallized into the figure of the fabled Tithonus as the classic representative of protracted age.[1] It is only in fable, however, that Tithonus suffers from the burden of undying age; in real life his frailties promote his euthanasia. Worn with his weary tramp through life, no longer able even to totter about, Tithonus at last lays him down to rest. Partly from the loss of the stimulating effect of the little exercise he was able to take, and partly from a similar cause to that which has occasioned the wasting of his skeletal muscles, his powers of assimilation give way. His blood becomes diminished in quantity and defective in quality; the brain centres for relative and for organic life get badly nourished; the genesis of force becomes more and more imperfect; slight wandering delirium sets in, and death from asthenia speedily follows. Scenes anticipatory of the future, more often memorial of the past, flit like dreams through the failing consciousness, and the weary mortal occasionally dismisses himself with some remark bearing on his future or his past. "Adsum"

In real life Tithonus dies. His death is a typical death from age.

---

[1] His wife, Aurora, obtained from Jupiter the gift of immortality for him, but forgot to ask for perpetual youth; hence, Horace says, "Longa Tithonum minuit senectus."— Lib. II., carmen xvi., l. 30.

has been the final utterance here, the fitting prelude to hereafter. Charles Abbot, the first Lord Tenterden, when dying, raised himself from his couch, and saying, with all his wonted solemnity, "Gentlemen of the jury," fell back and expired; and the gathering glooms of death drew from the great schoolmaster Adam the pathetic and appropriate farewell, "It grows dark, boys, you may go." "The great difference," says Bichat, "between death from old age and death from a sudden seizure, is that in the former death commences at the periphery and terminates at the heart — the empire of death begins at the circumference and ends at the centre; while in the latter death commences at the heart and spreads over the body generally — death begins at the centre of vitality, and gradually extends to its outmost bounds."[1] It is impossible to imagine any mode of dying to which Bichat's description of death from age could be more applicable than it is to that just described. It is the typical mode of dying from gradual decay. Except as to perpetual youth, Tithonus is no myth, and his mode of dying, though not the lot of every one, cannot fail to be recognized, and is not readily forgotten.[2] It is a mode of dying

---

[1] *Recherches physiologiques sur la vie et la mort*, Paris, 1805, p. 151.

[2] An admirable and most pathetic description of death from

peculiar to advanced age; yet, even in old age, men die more commonly from accident or disease than from simple decay. Not because development has not the same course in every one, and tends always to the same end; but because those tissue changes, which mark the progress of development from the cradle to the grave, intensify after middle life all the dangers of acute diseases, and by accentuating any latent organic weakness or structural defect, inherited or acquired, often cause those to die from age who have scarcely begun to think themselves old. Tithonus the aged succumbs at last from failure of the genesis of force. He dies from asthenia due to failure of oxidation following failure of assimilation, primarily induced by changes in the circulatory system. We cannot trace the changes in the capillaries and arteries beyond the vessels themselves: we know not the cause of these changes. But it is an advantage to know, and there is a general consensus of opinion on this, that the arterial system, which leads the van in the development of the body, is also that upon

Age must be measured by tissue change, and not by years.

age is to be found in the Book of Ecclesiastes, Chapter xii. The authors of the Revised Version have somewhat added to the pathos of this description by substituting the word "caper-berry" for "desire."

which the finger of decay is earliest laid.[1] By watching the development of this system and its relations to the heart and other organs, we are timeously warned, and are often able successfully to oppose the beginnings of evil. To recur to Bichat's simile, though we cannot prevent the sapping of the outworks, we can reinforce the citadel, and thus we are often able to postpone the ultimate surrender. True, we cannot hope in this way to provide an *Agerasia*, nor even to restore the patriarchal ages; but we can assuredly diminish the number and intensity of those side issues which so often bring life to a premature termination. We can greatly lessen human suffering, and we may put it in the power of many more nearly to attain the norm of life, which, according to Beneke, is from ninety to one hundred years.[2]

---

[1] *Vide* articles on "Age" in the *Cyclopedia of Practical Medicine*, p. 38, and in the *Cyclopedia of Anatomy and Physiology*, p. 77. *Vide* also Gimbert, "Memoire sur la structure et sur la texture des artères," *Journal de l'Anatomie*, Vol. ii., p. 648. Valerie Schiele-Wiegandt says: "In bezug auf das Alter ergiebt sich folgendes Gesetz sowohl bei Männern als auch bei Frauen nehmen im Grossen und Ganzen, entsprechend den höheren Altersperioden, in allen arterien umfang und dicke, respective media und intima, allmählich steigend zu" (*Virchow's Archiv.*, Bd. lxxxii., S. 36); and Roy has found that sometimes before, and certainly always after, middle life, the arteries begin to lose their elasticity (*Journal of Physiology*, Vol. iii., p. 125, etc.).

[2] *Op. cit.*, S. 26.

CHAPTER II

HOW THE HEART IS AFFECTED BY AGE

Two organs largely escape the effects of normal failure — the brain and the heart. Goethe, Von Humboldt, Leopold Ranke, Mrs. Somerville, and Thomas Carlyle, are memorable examples of those who have done excellent brain work at very advanced periods of life; and, indeed, the wisdom of age would never have become proverbial had the brain not been observed to functionate with its wonted integrity even in hoar age. In typical death from age the mind can scarcely be said ever to fail; it wavers, indeed, amid the gathering glooms of death, but till then its acuteness and energy are often scarcely diminished. The brain remains vigorous to the last, because its nutrition is specially provided for. At or after middle life, though the arteries of the body generally lose their elasticity and become slowly dilated, the internal carotids continue to retain their pristine

Heart and brain largely escape senile failure.

The maintenance of brain power specially provided for.

elasticity and calibre,[1] so that the blood pressure within the cerebral capillary (nutritive) area remains normally higher than within the capillary area of any other organ in the body; the cerebral blood paths are thus kept open, and the brain tissue itself is kept better nourished than the other tissues of the body.

The corollary from this is important: brain failure, not being a necessary characteristic of age, must always be looked upon as an indication of local malnutrition, and for this cardiac failure or arterial atheroma are most often to blame. In the one case, improvement may be expected from treatment; in the other, the failure of treatment but confirms the provisional diagnosis.

As for the heart, this organ has long been known to be hypertrophied in all old people. M. Bizot, in his well-known paper entitled "Recherches sur le Cœur et le Systeme Arteriel chez l'homme,"[2] tells us that "old age is in both

The heart is always found hypertrophied at advanced ages.

---

[1] Beneke, *op. cit.*, p. 24, "Die grossen Arteriellen Gefässtämme erfahren dagegen eine immer mehr zunehmende Erweiterung, und erreichen, *mit ausnahme der Carotides communes*, relativ zur korperlange eine noch betrachtlicher Weite, als im ersten Lebensjahre." Also, *op. cit.*, p. 75; and *Constitution und Constitutionelles Kranksein des Menschen*, von Dr. F. W. Beneke, Marburg, 1881, p. 42.

[2] *Vide Memoires de la Société Médicale d'Observation*, Tome premier, Paris, 1837, p. 262.

sexes that period in which the heart attains its greatest dimensions," so that if it be correct " to compare the size of the heart at thirty years of age to that of the fist of the subject, at sixty the heart will be found to be much larger if it is not abnormal." [1]

Charcot, in his "Lectures on Senile Diseases," [2] says that, unlike every other organ in the body but the kidney, the heart preserves even in old age the dimensions of middle life; and he adds that in some old people "the heart may even undergo a real hypertrophy."

Cohnheim is of a similar opinion; he says, " The heart of very old persons does not, as a rule, participate in the general atrophy of the body, and especially of the muscles, but rather increases in mass and volume." [3]

Beneke's experience, on the other hand, sufficed to convince him that only those reach advanced life who have been originally possessed of large and strong hearts.[4] According to Beneke age is

[1] " Si donc à trente ans le cœur doit avoir le volume du poing du sujet, à soixante il doit être plus volumineux, sous peine d'être dans une condition anormale. La vieillesse est, dans les deux sexes, l'epoque de la vie à laquelle le cœur offre le volume le plus considérable " (*op. cit.*, p. 275).

[2] New Sydenham Society's Translation, p. 28.

[3] *Lectures on General Pathology*, New Sydenham Society's Translation, Vol. i., p. 106.

[4] *Die Altersdisposition*, p. 24. " Wenn die von mir auf

not a possible inheritance of all, but only of a select few destined to it from birth. I myself, however, have seen too many weak hearts, and even hearts mechanically defective, attain advanced age to regard this idea as even approximately true.

Charcot distinctly recognizes the senile hypertrophy of the heart as the legitimate result of the senile alteration of the arteries. But he limits the change to "some old people," and regards it as pathological.[1]

Bizot, on the other hand, states explicitly that this senile hypertrophy of the heart occurs in all without exception; that it is mainly limited to the left ventricle, though the right ventricle also shares in it to a limited extent; and that it is invariably associated with dilatation of the arterial system and thickening of the arterial walls. These changes affect every one, man and woman alike, and continually increase as age advances.[2] But changes which happen to every one, and continually progress as age advances, form part of our development and are physiological, and not

Tab. I, gezeichnete curve (des Herzensvolum) in den 70ger Jahren noch wieder eine Hebung zeigt, so lässt dieselbe kaum eine andere Erklärung zu als dass diese hohe Altersstufe überhaupt nur von im allgemeinen kraftigen Naturen erreicht wird, und dass diese auch von Haus aus schon ein grosseres Herzvolum besitzen."

[1] *Loc. cit.*   [2] *Vide op. cit.*, pp. 275, 286, 288, 301, etc.

pathological. They may often enough be associated with pathological alterations of the arterial coats, but this is not always the case even at the most advanced ages, and the merely normal loss of arterial elasticity is quite sufficient to account for the change in the structure of the heart.

The normal elasticity of the arterial coats converts the intermittent blood flow from the heart into a continuous flow into the capillaries, and when this elasticity fails, the outflow into the capillaries becomes approximately intermittent, the blood pressure within their area falls, many of them obsolesce, and the most obvious, if not quite the earliest, of our senile changes are initiated. On the other hand, this intermittent outflow from the arteries accumulates the blood within them, and raises the intra-arterial blood pressure.[1] The result of this is that the left ventricle is called upon for extra exertion, in order to bring the arterial and venous blood pressures and the velocity of the circulation to their normal values. *The sources of vigour in the senile heart.* Fortunately the heart always works so well within its powers, that in health it readily responds to any call of this character.[2] The response of the left ventricle to

[1] *Vide antea*, p. 15.
[2] *Vide* Balfour's *Clinical Lectures on Diseases of the Heart and Aorta*, Churchill, London, 1882, second edition, p. 84, and p. 137, note.

this call is necessarily followed by the flushing of the myocardium at each pulsation with blood at a pressure considerably above the normal, hence — other things being equal — metabolism is more complete and nutrition more perfect. Add to this that according to Leichtenstern[1] the hæmoglobin is always found to be increased after sixty, and we see that the conditions at and after middle life are — in health — most favourable for the gradual development of hypertrophy of the heart, and especially of the left ventricle. Nay, so favourable are those conditions that weak hearts, and even hearts mechanically defective, are able to profit by them, so that many hearts at seventy are stronger and better fitted for the discharge of their functions than they were at sixty. The changes in the arteries due to age proceed slowly, imperceptibly, and so far as the individual himself is concerned, unconsciously. If the heart responds normally to the call for extra exertion demanded of it, the individual gradually descends into the vale of years quite unconscious whether he has a heart or not. If this knowledge is forced upon him, trouble is not far off.

Senile vascular changes proceed insensibly.

Various circumstances may bring to mind that

---

[1] *Untersuchungen über den Hæmoglobingehalt des Blutes in gesunden and kranken Zuständen*, Leipzig, 1878, S. 29.

we have a heart; its function may be disturbed by an excessive strain thrown on the myocardium by an early and excessive development of arteriosclerosis, the arteries in early life being sometimes as hard and tortuous as they are ever found to be even at the most advanced age. Ventricular embarrassment is produced by whatever increases peripheral resistance. *Causes of trouble to the senile heart.* We have, therefore, to reckon not only with the alterations in the elasticity and structure of the arteries, but also with the permanent contraction of the vascular area due to capillary obsolescence, as well as with those temporary contractions arising from reflex causes of various origins, which not only embarrass the circulation, but also give rise to sundry symptoms of very serious import. Moreover, peripheral resistance is greatly increased by any augmentation of the quantity of the blood, whether that be caused by plethora or hydræmia, and, as we can readily understand, it may be notably affected by the condition of the vascular environment.[1]

In estimating the various causes which hinder the passage of the blood from the arteries to the

[1] *Vide Text-book of Pathology*, by D. J. Hamilton, M.B., etc., London, 1889, Vol. i., p. 630 and p. 694. Donders seems to have been the first to direct attention to the importance of the vascular environment in relation to blood pressure. *Vide Physiologie des Menschen*, Leipzig, 1856, Vol. i., S. 169.

veins, and thus increase the intra-arterial blood pressure, we are too apt to overlook the condition of the tissues generally. We figure to ourselves the blood going its round through arteries, capillaries, and veins, as it were through naked tubes, forgetting that nutrition is extra-vascular, that the tissues are always flooded with blood-plasma, and that this fluid diffuses the elastic pressure of the tissues, and binds it to that of the arterial wall. The tissues themselves lose their elasticity through age, — like the arteries, — and this cannot be renewed. But the influence of the environment depends not so much upon this as upon the amount of fluid filling the interspaces of these tissues, and this varies both as to quantity and quality according to the state of the circulation, the quality of the blood, and the integrity of the secreting organs upon which this quality depends.

On the side of the heart embarrassment is brought about by all those circumstances and conditions of life which of themselves weaken that organ, and consequently intensify the action of those hindrances that have just been referred to.

Acute diseases weaken the heart by interfering with its nutrition and exhausting its nervous energy. Sudden critical or precritical cardiac collapse is a thing we are all well acquainted with. Sudden death from cardiac failure that not infrequently follows any abrupt exertion — such as sitting up or getting out of bed — during convalescence from acute disease, is also not unknown. But there is a third mode of dying from the heart after acute disease, which is neither so common nor so generally recognized; in this some trifling exertion, undertaken before the heart has had time to reaccumulate sufficient energy, starts an ingravescent asthenia from which there is no recovery.

Death from the heart, in any of those modes, is naturally most apt to occur after middle life, first because the cardiac energy is then more readily exhausted, and second because its action is already embarrassed, by one or more of the causes of peripheral resistance just alluded to.

Apart from acute disease, which, as we see, is more apt to initiate death, rather than heart trouble, chronic disease has an influence in this direction, but chiefly those forms of it which weaken the myocardium or impoverish the blood without materially diminishing the amount of the circulating fluid. Long-continued dyspepsia is well known as a common cause of heart trouble; sometimes it is only a symptom, but often it is a cause as well.

Loss of blood from any cause, either sudden and considerable or more continuous and in less amount, weakens the myocardium and leads to heart trouble, which worsens as hydræmia is established. Any other discharge which has a similar effect is followed by a similar result. Sexual excess has an equally ill effect, probably quite as much from loss of nervous energy as from any drain on the system.

Over-indulgence in food induces plethora — a most dangerous condition for any one with a weak heart. Plethora produces corpulency and loads the tissues with fat; this weakens their structure, and by making the cardiac muscle less fit for its function, it intensifies the action of the peripheral obstruction it helps to cause in inducing heart trouble. The abuse of stimulants and narcotics is a most fruitful source of senile heart trouble, and when conjoined with gluttony the combination is the most potent source of heart trouble we could have.

Sudden, violent, or unduly prolonged exertion is a fruitful source of heart trouble at all ages, but it acts with tenfold efficacy after middle life, and is a not infrequent cause of an abrupt termination to life itself.[1]

---

[1] The influence of overwork and strain in producing cardiac dilatation has long been known, and has been well described by Dr. Thomas Clifford Allbutt in Vol. v. of *St. George's Hospital Reports*, 1870; Dr. J. M. Da Costa, *American Journal of Medical Sciences*, January, 1871, p. 17; A. B. R. Myers,

Lastly, emotion of every kind has long been recognized as having an important influence on the heart's action and functions, and as a factor we dare not neglect in investigating the etiology of cardiac disease, and especially of sudden cardiac failure.

Inhibition of the heart's action by violent emotion is a well-known though unusual cause of sudden death; and contrary to what one would expect, joyous emotions are much more fatal than grief or sorrow.[1]

But such a tragedy as this is infinitely rare in comparison with the pathetic manner in which life is every day shortened by the petty troubles, anxieties, and worries which are of daily occurrence. The less intense but more persistent emotion keeps up a continual inhibition of the heart's action in a lesser degree. This impairs the ventricular systole, and coupled with those vascular conditions which after middle life favour cardiac dilatation, often precipitates heart trouble in those

---

surgeon, in *Etiology and Prevalence of Disease of the Heart among Soldiers*, London, 1870. Also in *Zur Lehre von der Ueberanstrengung des Herzens*, von Johannes Seitz, M.D., Berlin, 1875; and in *Die Herzkrankheiten in Folge von Ueberanstrengung*, von E. Leyden, Berlin, 1886.

[1] For many instances of sudden death from emotion, *vide A Treatise on Experience in Physic*, London, 1772, Vol. ii., p. 268. This is an anonymous translation of a work by Johannes Georgius Zimmermann.

who might otherwise have escaped. There are few of us who have been in practice for even but a short time who have not had occasion to note the development of serious cardiac symptoms from the trouble arising out of untoward domestic affairs, the worry of an unsuccessful business, or even the wear and tear of a too successful business which has outgrown the physical powers of its manager.[1]

The morbid anatomist finds after death, and ascribes to senile degeneration, many conditions, such as pigmentary involution, fatty degeneration, aneurism, and rupture of the heart.[2] But none

---

[1] E. Leyden, *op. cit.*, p. 47, says: "Die alten Aerzte wussten es sehr wohl, das Gemüthsbewegungen und Leidenschaften, Zorn, besonders aber Gram und Schmerz Herzkrankheiten zu erzeugen im Stande sind. Die neuere Medizin hat diese Erfahrung ziemlich vernachlässigt, doch wird jeder erfahrener Praktiker Beispiele davon anführen können. Ich selbst habe eine Anzahl solcher Fälle beobachtet. Sie haben in ihren Symptomen und ihren Verlaufe eine auffalige Uebereinstimmung mit den Fällen von Körperlicher Ueberanstrengung des Herzens und Man könnte versucht sein sie als psychische Ueberanstrengung jenen an die Seite zu stellen. Die Analogie besteht sowohl darin, dass das auffälligste Symptom, nähmlich die Arythmie und die Herzdilatation sich ebenfalls in Folge von psychischen Einflüssen entwickeln, als auch darin, dass zwei stadien der Krankheit unterscheiden werden können, das erste der Herzerithismus, das zweite die organische Herzdilatation."

[2] *Vide* Hamilton, *op. cit.*, pp. 582, 587, etc. Sclerosis and waxy degeneration are sometimes reckoned as senile changes, but the one is the result of inflammation, and the other of a

of these morbid states have any pathognomonic symptoms; it is only when they affect the heart's action that they come under the cognizance of the physician. It is much the same with the senile heart; its essential lesion is a weakened myocardium, rarely without dilatation of the cavities. This dilatation is caused by overstrain, occasionally from actual over-exertion, but far more frequently slowly induced by causes which are partly physical and partly nervous in their origin.

Essential lesion of the senile heart.

The symptoms of this weakened myocardium vary somewhat in each case, but they have a generic similarity in all. Precordial anxiety is usually what is first complained of, and however indefinite this may sound it is a source of extreme distress to the patient. Breathlessness, pain, or cardiac irregularity, in one or other of its many forms, are also early symptoms, and sometimes the case is accentuated by the conjunction of two or more of these symptoms.

Symptoms of the senile heart.

As these symptoms may all be present in the

general blood disorder; and both may occur at any age. Hamilton, *op. cit.*, p. 588, etc. *Vide* also *Etude sur le Cœur Senile*, par le Dr. Ernesto Odriozola. Paris, 1888. Huber, however, inclines to reckon sclerosis of the myocardium as a purely senile disease, seeking its origin in arterio-sclerosis alone. *Vide Archiv f. Patholog. Anatomie*, Bd. lxxxix., 1882, S. 236.

absence of any definite signs of any cardiac lesion, they are often grouped under the somewhat indefinite term of a NERVOUS HEART. A term applicable enough if only employed to signify that these symptoms are brought about through the agency of the nervous system, but quite incorrect if employed to suggest that these symptoms have no basis of physical change in the heart itself.

In its later stages the senile heart, in one of its forms at least, is the LUXUS HERZ of German authors, the GOUTY HEART of our writers. The term "gouty heart" is indeed equally applicable to its early as well as to its later stages, inasmuch as those vascular changes which superadd the gouty element proceed *pari passu* with those which originate the senile heart itself, and are closely linked with them. It is convenient, moreover, to have such a term to apply, as few people object to be called gouty, though many resent being called either nervous or old.

CHAPTER III

THE SYMPTOMS AND SIGNS OF THE SENILE HEART

It may be accepted as an axiom that all cardiac symptoms complained of after middle life, that cannot be distinctly referred to some evident disease, or to some affection of the cardiac mechanism due to disease, may be regarded as originating in actual or relative weakness of the myocardium. These symptoms may be of the most varied character. *Actual or relative weakness of the myocardium the origin of the symptoms of the senile heart.*

The heart working easily within its powers has its work — actual or relative — gradually increased, till it reaches a point when it makes itself felt. No longer unconscious of the existence of a heart, the individual becomes uneasily cognizant of the presence of that organ. The earliest indication of this is a feeling of emptiness and uneasiness in the left chest, very aptly expressed by the term *Precordial anxiety*. *Definition of precordial anxiety.*

If we examine the heart at this stage, we find

on palpation that any sensation of pulsation in the cardiac area is but feeble, while the apex beat itself is at the best weak and may be quite imperceptible; the percussion dulness is normal; on auscultation the sounds are normal, or, if there is any change at all, the aortic second is accentuated. These are indications of weakness of the myocardium, and the accentuated aortic second, if present, is an indication that to the normal loss of arterial elasticity there has been superadded a dilatation of the ascending part of the aorta.[1] Being weak, the heart is erethistic, it is irritable, and its action is readily excited or deranged by exertion, or by emotion, or by any other cause of reflex disturbance of the innervation. Hence to precordial anxiety we have superadded at least occasional irregularity of the heart's action in relation to rate, force, and rhythm.

Signs to be found at this stage.

The heart beats because its muscular fibre is incompletely differentiated, and still retains the power of spontaneous movement possessed by all primordial protoplasm.[2] The heart's energy resides in its muscular fibre, and its quality depends upon the perfection of the cardiac metabolism.

Movements of the heart primordial in character.

[1] Balfour, *op. cit.*, p. 31.
[2] Foster's *Text-book of Physiology*, 5th edition, 1888, p. 288 *et antea*.

The nervous system neither initiates nor maintains the rhythmic movements of the heart, but it controls and regulates them, and through it these movements may be variously modified and even arrested. *Influence of the nervous system on the cardiac movements.*

The agency by which the cardiac movements are controlled consists of a network of nervous filaments covering the surface of the heart, particularly at its base. On the one side, this network is connected with various nervous ganglia, scattered throughout the substance of the heart, particularly at the junction of the *sinus venosus* with the auricle, and in the auriculo-ventricular sulcus. On the other side this network unites into three distinct nervous cords, each of which plays a special part in regulating the movements of the heart. One of these cords (*H*, Fig. 1) passes through the first dorsal and the last cervical ganglion into the sympathetic nerve, and through it there pass to the heart those impulses which increase the rate of its pulsations and augment their force.[1] Acceleration and augmentation of the pulsations are not, however, necessarily coincident.[2]

[1] *Untersuchungen ueber die Innervation des Herzens*, von Albert v. Bezold, Leipzig, 1863, erste Abtheilung, S. 162.

[2] "Sometimes the one result, and sometimes the other being the more prominent." — Foster, *op. cit.*, p. 294. *Vide* also Roy and Adami, *Transactions of the Royal Society*, Vol. 183, p. 240.

At times we have a rapid heart-beat with a quick, large, and full pulse, but at other times the heart-beat is rapid while the pulse remains small (*vide Tremor cordis* and Tachycardia, *postea*). Further, it is through this nerve that the cardiac metabolism is effected, and its energy set free — it is the *Katabolic* nerve of the heart.[1]

The nerves of the heart and their actions. The other two nervous cords which pass from the cardiac plexus (*G* and *F*, Fig. 1), both enter and ascend to the brain along with the vagus nerve (*E*, Fig. 1), but each has its separate origin and function. The superior cardiac nerve (*F*, Fig. 1) is an afferent nerve, and conveys from the heart a controlling influence to the vasomotor centre in the *medulla oblongata* that regulates the movements of the arterioles, so that when a heart is labouring against a blood pressure too high for its powers, an impulse from the heart to this centre inhibits

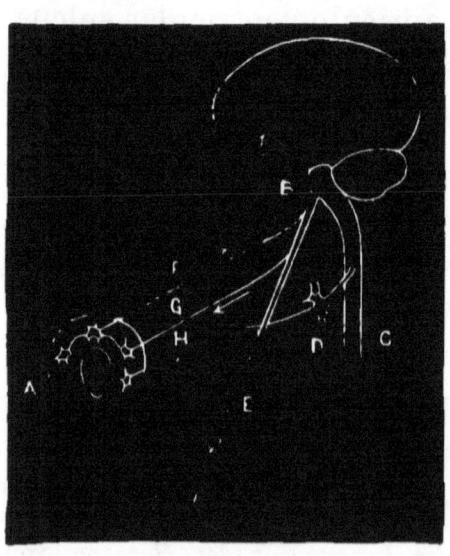

Fig. 1.

[1] Gaskell, *Journal of Physiology*, Vol. vii., pp. 41 and 46.

the constrictor influences and tempers down the blood pressure to suit the cardiac strength.[1] It is often called the Depressor Nerve of the heart.

The inferior cardiac nerve (*G*, Fig. 1), though it leaves the chest in the vagus bundle and is always referred to as a branch of the vagus nerve and its action as vagus action, is really more closely connected with the spinal accessory than with the vagus proper, and has a distinct root of its own. Von Bezold believed that this nerve was in constant action and thus supplies the natural tonicity to the heart.[2] In the present day this tonicity seems rather to be regarded as the property of the cardiac muscular fibre and to depend upon the perfection of its metabolism. The action of the vagus is *Anabolic;* it inhibits the action of the augmentor or katabolic nerve, it slows and reduces the force of the auricular action, and may even wholly arrest it for hours. On the ventricles the vagus has not so powerful an effect; strong stimulation of the vagus may indeed arrest the action of the ventricles, but never for a period long enough to endanger life. When the vagus excitation reaches a certain degree (varying in different animals), the ventricles begin to beat independently of the sinus and the auricles, and this idio-ventricular action, at first slow and irreg-

[1] Foster, *op. cit.*, p. 351.   [2] *Op. cit.*, S. 84.

ular, gradually becomes fairly rapid and almost completely regular in its rhythm. The interference of the sinus and ventricular rhythms with each other is the usual cause of irregularity of the heart's action, though irregularity may also be brought about by the auricles not responding to all the impulses which reach them from the sinus.[1] Irregularity of action diminishes efficiency of the heart without reducing its expenditure of energy. This unfavourable effect is of little consequence if the intermissions are infrequent, and the heart has a good margin of reserve force, but when the irregularity is great, or the heart feeble, diseased, or otherwise handicapped, it may form a very serious element of danger.[2] In no class of cases is this danger more likely to be serious than in cases of senile heart, because in these all the elements of danger act in combination towards one result — dilatation of the ventricles, more especially of the left ventricle. The elements of danger in such cases are: first, the normal, and still more any abnormal, increase in the aortic blood pressure; and, second, any abnormal diminution of the force of the

The cause of cardiac irregularity.

Danger of cardiac irregularity.

Why most serious in cases of senile heart.

---

[1] Roy and Adami, *op. cit.*, pp. 293, 294, etc.
[2] *Op. cit.*, p. 284.

ventricular contractions from malnutrition, or otherwise. Each of these conditions prevents the ventricle from emptying itself, and increases the amount of residual blood in the heart, and then there comes into action the law that "the strain upon the walls of a sphere or spheroid increases with its circumference, and, therefore, the resistance to contraction of the heart wall is increased whenever it becomes dilated."[1]

Vagus action slows the heart generally, lessens the excitability of the ventricles, and even when weak may reduce the output from them by as much as thirty per cent,[2] thus causing residual accumulation and all the evils that flow from it. But inhibitory influences may pass to the cardio-inhibitory centre from every quarter, and we can thus understand how mental worries or even physical derangements may, by a long continuance of petty inhibitions, seriously affect the cardiac function, especially when these inhibitions occur at a period of life when normal alterations in structure tend to accentuate their evil influence.

Irregularity or intermission should, therefore, never be looked upon as unimportant; as a drop hollows a stone *non vi, sed sæpe cadendo*, so even a simple intermission may ultimately lead to cardiac

---

[1] Roy and Adami, *op. cit.*, p. 213.
[2] *Op. cit.*, p. 217.

dilatation and the shortening of life. I well remember an old gentleman who for long complained of what he called an occasional "dunt" in his chest. This "dunt,"—throb,—which was his only complaint, was nothing but the augmentor impulse following an inhibition. When I first saw him his heart was considerably dilated, and had been so for some time, as I learned from his medical attendant. But the old gentleman was quite distinct in his statement that many years previously, when seen by a distinguished consultant in the west of Scotland, the doctor had told him that his heart presented no sign of disease, but only of nervous derangement. As his complaint had been all along the same, I have no doubt that the intermissions existed then, but that they had not as yet produced that dilatation which subsequently resulted. There was a strong suspicion that this patient indulged in some narcotic, but this was never brought clearly home to him, and he never confessed. He was somewhat relieved by treatment, but there was no marked improvement, and he was found dead in bed not long subsequently, no other symptom having developed. Simple intermission was in this case the earliest and the most persistent symptom, and it must have had a most important effect upon the course of the disease.

Illustrative case.

Intermission is a reflex inhibition of the heart through the vagus, of little consequence in youth, because the heart has then a wide margin of reserve; but of serious import after middle life, because all the conditions then present accentuate the tendency of persistent intermissions to induce dilatation of the heart.

Danger of intermissions.

The inhibitory cause may be of any character and may come from any quarter; it may be physical or emotional, a diseased organ, a depraved secretion, or a mental shock. A violent emotion — more especially, strange to say, if it be a joyous one — may fatally inhibit the heart's action; a less violent but more persistent inhibition lessens the ventricular output, increases the residual accumulation, and ends by dilating the heart; and if the exciting cause act with intensity, and the heart is already enfeebled, the dilatation may be rapid and acute.

Inhibitions may be physical or emotional.

The shock of a railway accident has been known to inhibit even a strong heart, and cause it to intermit; but this is of little consequence in a young heart with a good margin of reserve, as the effect ultimately dies away. I have known the intermissions from this

In a strong heart the effect of even a powerful inhibition ultimately dies off.

cause to drop within six months from one in every two beats to one in every twenty, and I have no doubt that they ultimately ceased altogether.

On the other hand, I have known the shock of a by no means serious railway accident so to break down a commencing senile heart as to make within a year an infirm and dropsical invalid of an active business man, and to kill within eighteen months a man who up to the time of the accident had scarcely been known to ail. Yet my own experience enables me to say with considerable certainty that we may reckon from twelve to thirteen years as the time required for the development of serious dilatation in a middle-aged man leading a life of ordinary activity, but without hard work, and taking no special care, but also having no special worries, reckoning from the time the patient was first led to consult a physician on account of cardiac symptoms which were then regarded as unimportant. The time mentioned is, however, merely approximate, and though based upon observation, it is liable to many modifications the sources of which are obvious.

I have spoken of a railway accident as a probable cause of cardiac inhibition, because in these days

Even a trifling inhibition may prove rapidly fatal to a weak heart.

Time required to develop dilatation reckoning from the earliest symptom.

such an accident is one of the most common causes of serious shock; but other forms of accident may be equally injurious, the result depending much more — in regard to cardiac disturbance — upon the violence of the attendant emotion than upon that of the physical shock. Thus Richardson narrates a case of shipwreck in which the fear of instant death from drowning caused the heart of a middle-aged man, in perfect health and spirits, suddenly to stop. He was rescued from his sinking ship and put on board another vessel, and when he had regained sufficient composure he found that his heart intermitted four or five times a minute. At first these intermissions were so disturbing as to prevent sleep; by and by they died away to two in a minute, and the patient was no longer cognizant of them unless he felt his pulse.[1] I have no doubt they ultimately ceased entirely, but this is not recorded. Richardson also tells us of another case in which wearing anxiety of purely mental origin developed persistent intermittency, followed by death from the silent but sleepless suffering produced,[2] cardiac dilatation having doubtless an important influence on the fatal issue.

Intensity of emotion an important factor.

---

[1] *Transactions of the St. Andrews Medical Graduates' Association*, 1870, p. 238.   [2] *Op. cit.*, p. 239.

We are all well acquainted with the intermissions due to the gastric irritation arising from flatulence, undigested food, or other disturbance, which are more prone to affect those with long, narrow chests than those with roomier paunches. We know also those intermissions due to the abuse of alcohol, tobacco, or other similar poisons. Such reflex or direct inhibitions are fortunately more easily remedied than many, but they are no less injurious to a senile heart, and they require to be carefully attended to and their recurrence prevented.

The following sphygmogram (Fig. 2) was taken from the radial artery of a man whose feeble and

Fig. 2.

irregular pulse was the cause of considerable anxiety to himself as well as to his medical attendant, especially as there was no very evident reason to account for it. I found this patient to be a man of regular and unimpeachable habits, but that his health was considerably below par, apparently from confinement during office hours to a badly ventilated apartment in which a great deal of gas was burned. He was a valuable servant, and his employers were quite inclined to do their best for him; so I told him to get his office properly ventilated, giving him at the same time a tonic mixture. The result was most satisfactory — within a month his health was quite re-established and his heart steadied; he has kept well ever since, and conducts successfully the business of a large and important company.

Here we had a young man (æt. 36), organically quite sound, yet his life made useless and miserable by a heart feeble, intermitting, and irregular, because the blood in its coronaries was impoverished and depraved — a very good example of the apparent effect of the quality of the blood passing through the coronary arteries in governing the heart-beat.[1]

Chlorosis and Anæmia, especially that form of it — Hydræmia — where the blood is plentiful enough but of poor quality, are very common causes of this form of irregularity. At rest such patients have feeble but regular pulses, but the slightest exertion produces one of two things — either a rapid and forcible heart-beat, or marked irregularity of the heart's action. To maintain the perfection of the muscular metabolism, imperfectly oxygenated or otherwise impoverished blood has to be sent through the heart and other muscles

[1] Foster, *op. cit.*, 1891, p. 344.

much oftener per minute than healthy blood; the augmentor nerve is therefore called into action, and the heart-beat becomes rapid and forcible. Should the heart be fairly well nourished, the blood not much below the average in quality, and the exertion of but short duration, this is all that happens: the rapid and forcible heart-beat quiets down when the need for it ceases, and the heart is none the worse for its effort. But when the heart is not so well nourished, the blood more depraved, or the exertion more sustained, the katabolic action of the augmentor nerve becomes dangerous to the integrity of the organ as well as to that of the organism, and the anabolic action of the vagus is called into play. The suffering heart sends a message to the inhibitory centre, and the reply comes through the vagus as an inhibition which weakens the force of the auricular contractions, lessens the strength of the rhythmic excitation which reaches the ventricle from above, and at the same time diminishes the excitability of the ventricle itself.

How vagus interference causes irregularity.

If the need of the heart be urgent, the inhibition is strong, any stimulus reaching the ventricle from the auricle is but feeble, and, consequently, the ventricle sets up its own independent rhythmic action.[1] The auricular rhythm and the inde-

[1] *Vide antea*, p. 40, and Roy and Adami, *op. cit.*, p. 293, etc.

pendent ventricular rhythm are each quite regular in themselves, but when they affect the ventricle at the same time, they interfere with each other and set up arhythmic irregularity.[1] The following sphygmograms are examples of this irregularity as occurring in feeble and dilated hearts.

The one (Fig. 3) is from the radial artery of a dignitary of a southern university, who died within a year from the time this was taken; the other (Fig. 4) is from the radial of a man who still survives and is well.

Fig. 3.

When the heart is weak and dilated, or the blood much impoverished, the most trifling exertion

[1] Roy and Adami, *op. cit.*, p. 283.

may call for anabolic action and give rise to arhythmic irregularity. But when heart and blood are only slightly below par, the violent exertion of a foot-ball match may be needed to induce irregularity, and under the restorative influence of rest this speedily dies away.

Very important information as to the condition of the myocardium, and the state of the blood, is thus to be obtained from the greater or less readiness with which irregularity is evoked. The importance of this in the prognosis of a weak heart need scarcely be pointed out, and the value of a due recognition of the cause of irregularity in relation to the treatment of such a heart must be obvious to all.

Irregularity as to force is a common accompaniment of arhythmic irregularity, because every now and then the impulse from the auricle coincides with the ventricular systole, and there is an unusually full beat, readily recognized by the finger on the pulse, and just as easily seen on the sphygmogram. This marked irregularity of the pulse-force constitutes a very distinctive difference between the irregularity of cardiac dilatation and that which so frequently accompanies mitral stenosis. In the former case a certain number of the radial pulsations are full and large, while in the latter,

Cause of irregularity in the force of the pulse.

though the pulses do vary in force, there are none that can be called full or large.

Irregularity in rate, though not peculiar to the senile heart, is yet, in some of its most remarkable varieties, most commonly found associated with it.

These irregularities in the pulse-rate are interesting and important enough to require a separate chapter to themselves. I shall only mention them at present. *Varieties in the pulse-rate.*

Tremor cordis is a most remarkable phenomenon even to those well acquainted with it; and it is scarcely possible to conceive anything more alarming than a first attack of what Sir Walter Scott called the *morbus eruditorum*, but which, alas! is not nowadays confined to the erudite any more than *Podagra* is to the great and noble, with whom Sydenham flattered himself he had a community of suffering, and there is no one who suffers from this tremor who is not ready to exclaim with Sir Walter: "What a detestable feeling this fluttering of the heart is!"[1]

Tachycardia is a new name for an old complaint which, with its converse, *Bradycardia*, requires full detail to make it either interesting or under-

[1] *Vide The Journal of Sir Walter Scott*, Vol. i., p. 153. Douglas, Edinburgh, 1890.

standable. None of these irregularities in the pulse-rate are strictly limited to the latter half of life, but they are most common and most distressing then, and it is thus convenient, if not strictly accurate, to treat them as affections of the senile heart.

The senile heart as I have described it is essentially a heart that has been overstrained through inability to do its work, while in many cases, as just pointed out, this overstrain or dilatation is precipitated by nerve interferences. The degree to which the cardiac cavities have already yielded we learn mainly from palpation and auscultation. In the hands of an expert careful percussion is capable of yielding very trustworthy results in determining even a trifling increase of the heart's dulness, but this is difficult even in a male chest, and in a female one it is still more so. On the other hand, the results of palpation are easily obtained and readily comprehended; for example, when an individual over middle life has his arteries atherosed, an accentuated aortic second, a firm, tense pulse, or a sphygmogram indicative of high intra-arterial blood pressure, with his apex apparently beating in the normal position, palpation at once reveals whether this apparent apex-beat is really the outward thrust of the point of the left ventricle, or

Mode of detecting cardiac dilatation.

merely the edge of the right ventricle. For the true apex of a normal heart is the strongest point of pulsation in the cardiac area, whereas the apex in a dilated heart is felt to be a mere extension from the strongest point of pulsation lying beneath the lower end of the sternum. At even an earlier stage, long before the right ventricle has become so markedly dilated as to produce a pulsation beneath the sternum marked enough to be detected by palpation, the ear through the stethoscope can readily distinguish the abnormal strength of the right ventricular impulse.

The pulmonary circulation being a closed circuit, whatever hinders the onward flow of the blood through the left heart, exerts an equally obstructive influence on the flow through the right ventricle; this consequently dilates, and, as the right ventricle lies between the sternum and the left ventricle, a very slight dilatation suffices to push the left apex from the chest-wall into the cavity of the thorax, its place being taken by the right apex.

As the heart dilates, the apex-beat extends gradually outwards to the left till it reaches the nipple line, or even beyond it, keeping in the fifth interspace, but beating three or more inches from mid-sternum instead of only two and a half.

During the gradual dilatation of the heart its normal sounds undergo a progressive alteration that ends in a loud systolic murmur in all the cardiac areas. The sequence is as follows:—

Changes which the heart's sounds undergo during dilatation.

In the very earliest stage, when the sole symptoms are precordial anxiety, a feeble impulse, and an accentuated second sound (*vide* p. 35), the first sound is always more or less altered. It may be prolonged, blunt, feeble, or impure; now and then we have it loud, clear, and booming; over the right apex, when the heart is dilated, the first sound is always more distinct than over the left apex, except the booming sound, which is a left-side phenomenon, and best heard just below and to the left of the nipple. Distinctness of sound is probably as much due to the resonating qualities (thinness and flexibility) of the chest-wall as to any particular state of the ventricle; the booming quality conveys to the mind the idea of tension, and seems to indicate considerable dilatation of the ventricular cavity.[1] Often accompanying one or other of these alterations of the first sound, and certainly speedily following it

---

[1] For various views as to the state of the first sound in dilatation of the heart, *vide* Hope, *Diseases of the Heart*, 3d ed., p. 68 *et seq.;* Walshe, *Diseases of the Heart*, 3d ed., p. 315; Stokes, *Diseases of the Heart*, etc., p. 217. Also Gairdner, *Ed. Medical Journal*, July, 1856, p. 55.

in orderly sequence, the educated ear has no difficulty in detecting a systolic murmur in the auricular area, in appropriate cases. This auricular murmur is a murmur audible between the second and third ribs to the left of the sternum, just outside the pulmonary area. The pulmonary artery, as we know, comes to the front between the second and third ribs on the left side, one half of its breadth lying beneath the sternum, and the other in the interspace. If we put a finger-tip in this second interspace, close to the edge of the sternum, we cover the pulmonary artery, and just outside of the finger-tip the left auricular appendix in most hearts reaches the chest-wall, and if large and dilated, passes to the front of the ventricle at the root of the pulmonary artery. The *appendix auriculi* is not always long enough to reach the surface, and in such a case the auricular murmur is naturally not to be heard; but in all cases in which this murmur is audible it may be accepted as an early and infallible sign of mitral regurgitation,[1] and consequently of ventricular dilatation in cases such as those now referred to. Failing this auricular murmur, and often accompanying it,

Position and cause of the auricular murmur.

Why the auricular murmur is not always to be heard.

---

[1] Naunyn, *Berliner klinische Wochenschrift*, 1868, No. 17, S. 189; and Balfour, *op. cit.*, p. 171.

we have as an early sign of ventricular dilatation, an occasional systolic whiff over the apex. For a time this whiff is transitory, more audible at one time than at another, and sometimes entirely absent, replaced by a more or less altered first sound. These variations depend on the state of the circulation, the murmur being always most distinct after exertion and not so audible — often entirely absent — when the patient has been resting. But a systolic murmur, due to progressive dilatation of the left ventricle, does not long remain trifling or evanescent. Erelong it is to be found whenever listened for; it is speedily followed by a systolic tricuspid murmur, and then we have a systolic murmur in all the cardiac areas. In the aortic and pulmonary areas this murmur is partly due to propagation from the mitral and tricuspid openings, and is partly produced there, as an early phenomenon in the aorta, and a late one in the pulmonary artery, by the passage of the blood through the comparatively narrow arterial openings into the dilated arteries beyond. By the time the mitral and tricuspid murmurs have developed there is no difficulty in determining from its enlarged percussion area, and from its forcible impulse, that the heart has become, not only dilated, but also hypertrophied.

A transitory systolic whiff an early sign of dilatation.

How this systolic murmur spreads.

The accentuated aortic second is always an indication of a dilatable aorta, but by itself it is not a sign that the aorta is actually dilated. After death, in such cases, the aortic walls are always found to be inelastic; during life the aorta expands helplessly before the advancing blood-wave, which for want of its normal elasticity it fails to pass completely onwards. The excess of blood in the inelastic ascending aorta falls back on the sigmoid valves and closes them, with unusual force, by the mere virtue of its abnormal weight (or momentum). *What an accentuated aortic second indicates.*

At first, and for a time, the aortic second is merely accentuated in virtue of possessing a louder and more distinct sound than usual; by and by there is superadded a booming quality, which indicates closure by a heavier blood-column, throwing a greater tension on the aortic valve. As this tension gradually increases the segments of the valve tend to get separated, and to permit of regurgitation between them into the ventricle. In this condition anything which diminishes the size or weight of the aortic blood-column, or that increases the tone, or diminishes the extensibility of the aortic walls, diminishes the regurgitant force of the blood, and thus au *The meaning of a booming second.* *Curable aortic regurgitation.*

aortic regurgitation of this character is curable, and is occasionally cured.

Often, however, the aortic walls are not merely inelastic and dilatable, but rigid, atheromatous, and the lumen dilated. In these cases the blood-wave does not merely dilate the passive walls of an inelastic aorta, but passes through the relatively small aortic opening into the dilated artery beyond, and in so doing forms fluid veins which give rise to a systolic aortic murmur; an indication not of mere dilatability of the aortic walls, but of actual dilatation of the aortic lumen.

Cause of the systolic murmur in aortic dilatation.

A merely accentuated aortic second, then, only indicates with certainty an inelastic and dilatable condition of the aortic walls; but an accentuated aortic second, coupled with a systolic aortic murmur, indicates an actual dilatation of the aortic lumen, and this may be confirmed by percussing the aorta and mapping out its dulness. The history of the case, and the fact that diseased valves, capable of themselves — by obstructing the arterial exit — of originating a systolic murmur, are, from sheer inflexibility, incapable of accentuating the second sound, help to confirm the diagnosis. This actually dilated state of the aorta is, much more often than a merely dilatable one, followed by separation of the segments of the aortic valve, and by regurgitation into the ventricle. Hence it has long been known that in many cases — all, indeed, of this character — a systolic aortic murmur precedes for an indefinite period the development of regurgitation.[1] At first we have only an occasional diastolic whiff accompanying the booming second, and generally to be earliest heard at the sternal end of the fourth rib on the left side. As in the case of the systolic whiff in the mitral area, this diastolic aortic whiff gradually becomes permanent, and gets louder and more prolonged as the regurgitation becomes freer, until at last the case which commenced as one of precordial anxiety with an accentuated second, a feeble impulse, and an impure first sound, terminates as a *cor bovinum* with a heaving impulse, and a double murmur more or less audible in all the cardiac areas.

Precordial anxiety may terminate in a cor bovinum.

Erelong this condition is followed by renal congestion, albuminuria, and dropsy. Fortunately, all these troubles are preventable, and early attention to the beginnings of evil may not only avert these untoward results, but promote a green and healthy old age.

But this may be averted.

No arguments are required in the present day

---

[1] Stokes, *op. cit.*, p. 227.

to prove that stress of work, from increase of the intra-arterial blood pressure, is sufficient to induce dilatation of the heart. It is acknowledged by pathologists,[1] and has been experimentally induced by physiologists.[2]

Ventricular dilatation is speedily followed by regurgitation through the auriculo-ventricular opening, and this is now universally acknowledged to be accompanied by an impure first sound, which speedily develops into an unmistakable systolic murmur.[3] Various explanations have been given of this valvular incompetency without valvular lesion. The explanation which seems best to agree with the facts is that given by Ludolph Krehl. This observer points out, what has indeed been long known, that in the normal heart the valves are floated into apposition, and the auriculo-ventricular opening closed previous to the commencement of the ventricular systole;[4]

How regurgitation is brought about. Krehl's account of it.

---

[1] Ziegler's *Pathological Anatomy*, by Macalister, London, 1884, Part ii., p. 49.

[2] Roy and Adami, *British Medical Journal*, December, 1888, p. 1321, etc., and *Transactions of Royal Society, loc. cit.*, pp. 213 and 278, etc.

[3] The first recognition of this as a necessary complement of ventricular dilatation, and not a mere accidental complication, we owe to Dr. Gairdner. Vide "The Evolution of Cardiac Diagnosis," *Ed. Medical Journal*, June, 1887, p. 1080.

[4] *Vide* Pettigrew, "On the Relations, Structure, and Func-

were it otherwise, a manometer within the auricle would infallibly indicate regurgitation at the moment of systole. When, however, the ventricle is dilated, there is regurgitation — so-called relative insufficiency is established. Not because the auriculo-ventricular opening is dilated, — that is a later occurrence, — not because the segments of the valve are unable to close the opening, — one alone of these segments is almost sufficient for this purpose, — but because the insertions of the *chordæ tendineæ* into the papillary muscle, owing to the ventricular dilatation, are set so wide apart and so far from the centre of the ventricle that the trifling pressure of the auricular blood is unable to bring the valve-segments into apposition.[1] Under these circumstances whenever the ventricular systole commences, regurgitation occurs; at one time to but a limited amount, at another to a greater.

When we listen over the apex of a dilated heart, we hear in some cases only an impure first sound, in others a systolic whiff precedes a quite closed first sound, and in still others a murmur begins with the beginning and continues throughout the whole of the systole.

---

tion of the Valves of the Vascular System," *Transactions of the Royal Society of Edinburgh*, 1864, p. 799; and *Physiology of the Circulation*, London, 1874, p. 284.

[1] *Vide Archiv für Anatomie und Physiologie*, Leipzig, 1889, S. 291.

This agrees exactly with Krehl's account; when the ventricular systole begins, the valve-segments are not in apposition as they ought to be; but as the systole progresses all the conditions conduce to the perfect closure of the valve, when the dilatation is slight. It is quite a different story when the dilatation is considerable, or the valves diseased.

There are two other symptoms which are of serious import in advanced life, though neither of them is limited to that period.

As an accompaniment of imperfect circulation, pulmonary congestion, or defective hæmoglobin, *Breathlessness* is the commonest symptom of cardiac failure at every age; it is never absent when any exertion is called for. But connected with the senile heart breathlessness assumes a different aspect — exertion is not needed to induce it; it may occur when the sufferer is at perfect rest, and it may even awake him from sleep. When it harasses the patient in this way, breathlessness gets the name of cardiac asthma, and is often an early symptom of cardiac failure. In this connection it is a true *Angina*, and much more entitled — etymologically — to that appellation than the painful affection that commonly bears it. These two forms of angina will be treated of together, later on.

Two forms of Angina.

CHAPTER IV

PALPITATION, TREMOR CORDIS, TACHYCARDIA

PALPITATION is a common complaint of those who suffer, or who think they suffer, from disease of the heart. It is a term commonly applied to all forms of abnormal cardiac pulsations which make themselves unpleasantly sensible to the sufferer — to intermission as well as to irregular action. Therefore, although palpitation is not a symptom peculiar to the senile heart, — is, indeed, more apt to affect the young than the aged, — it is yet well to define clearly what is meant by this term, so that it may be differentiated from other forms of heart hurry. *What we mean by the term "palpitation."*

The distinctive peculiarities of palpitation are a regular, rapid, and violent pulsation of the heart, which often shakes the whole chest and always makes itself unpleasantly sensible to the sufferer, accompanied by a violent throbbing of the aorta, carotids, and other large arteries, which does not extend to the smaller vessels, the radial pulse giving no indication — in its force, at least — of the violence of the heart's action. Palpitation comes on suddenly, and may last from a few minutes to several hours; it is very distressing and often alarming to the sufferer, but it is not usually attended by any danger. It seems to be caused by reflex inhibition of the vagus action, a reflex paralysis of the inhibitory centre which removes the restraining influence of the vagus, and allows the augmentors temporarily to run off with the heart. Palpitation occurs in weakly and anæmic individuals, and is produced by reflexes of emotional or gastric origin, never by exercise. The rapid, forcible augmentor action that follows exertion in a spanæmic person (*vide* p. 47) simulates, indeed, palpitation closely; but in such a case the radials beat fully and forcibly, the heart's action is not so violent and throbbing, and all the phenomena cease at once whenever the patient becomes quiescent.

Tremor cordis is a very remarkable form of cardiac irregularity. It is the very opposite of palpitation. Emotion has nothing to do with its causation, and the heart, instead of throbbing as if it would burst the chest-wall, trembles like an aspen leaf. It occasionally occurs in youth; it is common enough in advanced life; it is most alarming not only from its peculiar character, but also from the sudden way in which it seizes its victims. We talk of a bolt from the blue as the most startling thing that could happen, but it could not be more startling than that a heart beating quietly and steadily should suddenly be seized with a rapid, tremulous fluttering, most alarming to the victim not only from the unusual character of the sensation, but also and especially because of the organ affected; for life truly seems slipping away when the heart itself trembles. This affection was well known to Senac and the early physicians, who seem to have taken rather a serious view of it. *Tremor cordis. What it is.*

These attacks occur without warning, and pass off in a few seconds, apparently without detriment to the patient. They are generally spoken of as "a fluttering of the heart," and such indeed they are. The sensation is precisely as if the gouty twittering of the muscles spoken of by Begbie[1] had affected the cardiac muscle. The pulse does not die away; it does not taper off like a *pulsus myurus*, but it suddenly drops from the ordinary full pulse of health to a mere tremulous thread. The attacks vary from three or four sharp, short, and apparently incomplete systoles, rapidly suc-

---

[1] *Contributions to Practical Medicine*, by James Begbie, M.D., Edinburgh, A. & C. Black, 1862, p. 6.

ceeding one another, and running off without warning from a heart beating regularly and steadily, up to a whole series of rapid, short, and incomplete systoles, which may last for several seconds, convey a tremulous sensation to the hand laid over the cardiac region, and are accompanied by a small, fluttering, and often scarcely perceptible, radial pulse. This *tremor* ends suddenly like an intermission, with an unusually forcible beat, and from a similar cause. During all those imperfect systoles, the ventricle has been getting gradually overfilled, the augmentor nerve is called into play, the ventricle forcibly expels its contents, which escape freely, and gradually distend arteries which have had time to get unusually empty. The heart then settles into its ordinary rhythm.

Tremor cordis is not confined to the senile heart. It may happen at any age. It may attack a heart apparently healthy, or it may accompany any form of heart affection; but it is most common after middle life, and in hearts which are feeble and dilatable. Sir Walter Scott called it the *morbus eruditorum*, and tells us that in his youth it used to throw him into "an involuntary passion of causeless tears."[1] I myself am well acquainted

---

[1] "I know," he says, "it is nothing organic, and that it is entirely nervous, but the sickening effects of it are dispiriting to a degree." — *Op. cit.*, p. 153.

with a man now getting on for seventy, who, at nineteen, was suddenly and without warning seized with a sharp attack of *tremor cordis*. This happened just previous to an attack of relapsing fever, and, up to quite recent times, it remained an only one. Of late, these seizures have been more frequent. His heart has always been irritable, but he has a long, narrow chest, and those having this conformation have almost invariably irritable hearts. He has always enjoyed good health, but may be said to be hereditarily disposed to heart affection on the mother's side. On the father's side, the deaths have been for generations, in the direct line, all over eighty, usually from cerebral apoplexy. Rheumatism is unknown on either side. Further, this patient tells me that occasionally he has been able to arrest this tremor by a voluntary impulse through the inhibitory centre, not absolutely, but markedly enough to his own sensation. Nor is this impossible. There is one medical man recorded by Fothergill as possessing the power of voluntarily arresting the heart's action,[1] and we are all acquainted with the remarkable case of Colonel Townsend, narrated by Dr. Cheyne;[2] not

Case of tremor cordis.

Voluntary arrest of the heart's action. How brought about.

[1] *Lancet*, I., 1872, p. 498.
[2] *The English Malady*, by George Cheyne, M.D., London, 1723.

to mention the Indian fakeers, who undoubtedly possess the power of arresting both pulsation and respiration, as we gather from those remarkable cases, narrated by Dr. Braid, where they submitted to be buried for so long as six weeks at a time, till their clothes were all mildewed and rotten.[1] This power is now believed to be exercised by voluntary compression of the spinal accessory by the muscles of the neck, which transmits a powerful inhibition to the heart through the cardiac branch of the vagus with which the spinal accessory is so intimately connected.

Tremor cordis, rare in youth, common enough after middle life, is always spoken of as a fluttering of the heart, and can generally be associated with flatulence or some other gastric disturbance. Never, in all my experience, has any form of emotion had any share in its causation. Indeed, it is the sudden way in which, without a thought being directed towards it, an apparently healthy heart, beating quite regularly and steadily, begins to flutter within the chest that makes it so alarming to the sufferer. No feeling of faintness seems ever to be connected with this most uncomfortable sensation. It is a most singular phenomenon, and difficult to explain satisfactorily. Evidently the

---

[1] *Observations on Trance*, by James Braid, M.R.C.S. Ed., A. & C. Black, Edinburgh, 1851.

vagus is reflexly inhibited, the heart uncontrolled goes off at a gallop, till the ventricle, which all the time has been gradually getting overfilled, suddenly invokes augmentor aid, expels its contents with a bang, and at once settles down steadily under normal nervous control. This explanation certainly agrees with the facts. The shorter the period of *tremor*, the less forcible the impulse with which the heart returns to work.

TACHYCARDIA, or heart hurry, is a symptom not confined to the senile heart nor to the latter half of life, but it is most dangerous to the aged, and in them it is always pathological. In infancy tachycardia is a physiological phenomenon, as the heart of the new-born babe beats at the rate of 130 per minute, gradually dropping to 100 at three years of age. In pathological tachycardia, the heart-rate is said to reach 200 or even 300 per minute. I myself cannot distinguish with any certainty over 150 pulsations a minute. By the aid of the sphygmograph we may certainly count more, but I have never found them over 200. One great distinguishing peculiarity of pathological tachycardia is the little disturbance it gives the sufferer. With a heart beating more rapidly than that of an infant, he goes about his duties as unconscious as a babe of anything unusual. This is one great difference between tachycardia and palpitation, with which it is so apt to be confounded.

During the first two years of life, the rapid action of the heart depends upon the low blood pressure, and concurs with it in promoting the diffusion of the blood-plasma and the rapid growth of the tissues. Infantile tachycardia is the necessary result of the conditions under which the circulation is then carried on; in its turn it is subservient to the building up of the frame, and it gradually ceases as the intra-arterial blood pressure rises and development takes the place of growth. At any later period of life tachycardia is an abnormal phenomenon, and indicates some interference with the physics of the circulation, or with those nervous connections by which its various interdependent relations are maintained and regulated.

Normal tachycardia.

In a few cases tachycardia is found in women at the menstrual period, or during the puerperium; it is also occasionally observed in both young and old recovering from an illness, their hearts never quite falling to the normal rate, and by some all of these varieties of heart hurry have been looked on as physiological. But all such cases are exceptional and essentially morbid in their causation, as are even those still rarer cases in which the rapid heart of infancy persists even to old age.

I am acquainted with one instance of this—a lady, now a widow over seventy, who has had a large family, and whose pulse up to quite recent years was never under 150 per minute; now it is only seventy. *Persistence of infantile tachycardia.* This lady is of a highly neurotic temperament, but she has always enjoyed good health, and there never was any violent or distressing throbbing either in the region of the heart or at the root of the neck.

In tachycardia the heart's action is rapid and feeble, and the sounds are empty, like the tic-tac of the fœtal heart, while the radial pulse is quick, feeble, and sometimes almost imperceptible—a state of matters by no means devoid of danger, and one which may terminate suddenly either in *Syncope* or *Asystole*, and which differs *toto cœlo* from other affections to which, so far as the heart-rate is concerned, the term tachycardia is equally applicable.

How completely, for example, does the Syndrome of such an affection differ from that of so notable an instance of rapid heart as exopthalmic goitre. *Syndrome of exopthalmic goitre.* And yet in Graves' disease there is often for months neither exopthalmos nor goitre—nothing but a rapid heart.

There is always tachycardia so far as rate is concerned, the pulse beating 140 or more per minute; but the heart's action is violent, and the whole arterial system throbs disagreeably. The heart sounds are clear and distinct, and sometimes so loud that Graves, to whom we owe the earliest description of this disease, says in reference to one of his cases, "I could distinctly hear the heart beating when my ear was distant at least four feet from the chest."[1] At times this violent perturbative palpitation of the heart and arteries exists alone; at other times this is associated with goitre only, or with exopthalmos only, or all these three symptoms may be present; but always and in every case the violent throbbing of the heart and arteries is sufficient to distinguish it from mere tachycardia.

Tachycardia always symptomatic. Those who believe in an essential tachycardia speak of the heart hurry of febrile or exhausting diseases as a symptomatic tachycardia; and in like manner the rapid pulse, feeble impulse, and empty heart sounds, which so often accompany and herald the approach of death, may with more reason be termed the *tachycardia morientium*, inasmuch as it is sometimes difficult to say whether the tachycardia is merely the herald or not also the cause of death.

[1] *A System of Clinical Medicine*, by Robert James Graves, M.D., Dublin, 1843, p. 674. This Lecture was first published in 1835.

For, while holding that tachycardia is only a symptom, it must still be acknowledged there are many cases in which it is the only detectable symptom. Are these to be considered cases of true, essential tachycardia or not? I feel certain that careful enquiry will in every such case discover some previous heart strain sufficient to originate an endocarditis or a myocarditis, some coexisting chronic disease, some history of an overwhelming emotion, or the abuse of some cardiac poison, any one of which may be quite sufficient to account for the predominant symptom. I have seen many cases of tachycardia due to heart strain; many of these got well without developing any further symptom. Whether these were cases of slight and evanescent endocarditis or of myocarditis no one could say. On the other hand, cases which throughout their whole course presented no other symptom have terminated in stenosis of the mitral opening, as revealed by all the ordinary and well-known signs, and have thus sufficiently plainly indicated their endocarditic origin. Never less than two years were required for the development of a presystolic murmur, reckoning from the first appearance of the tachycardia, and often much longer.

But tachycardia is not only, as it were, a cause of mitral stenosis; it is a very frequent accompaniment of that affection. Indeed, in a well-marked case of tachycardia, I would look first for the signs of mitral stenosis, and failing them, for the signs of a dilated heart with marked indications of arterial atheroma. In the one case the heart's action is apt to be not only quick, but also irregular; we have an accentuated first sound, and generally a well-marked pulmonary second, though sometimes from anæmia this is not so well marked as it ought to be; these signs, coupled with the history of the case, enable us to differentiate it from similar rapid hearts, with, at all events, considerable probability.

Tachycardia may terminate in mitral stenosis.

And is often an accompaniment of that affection.

On the other hand, when the heart is slightly enlarged, dilated, and hypertrophied, with persistent tachycardia, there is suspicion of interference with the coronary circulation, or of some condition of the blood involving imperfect metabolism of the myocardium.

May be caused by imperfect metabolism of the myocardium.

In the latter half of life tachycardia is a symptom associated with various forms of degeneration, and if not from the first dependent upon cardiac disease, it is always associated with cardiac dilatation. From almost the first, even in those cases which seem strictly essential, there is increased precordial dulness, dependent on imperfect ventricular systole with residual accumulation, which is so essential a part of the affection, and which is increased and accentuated by all those obstacles to the onward flow of the blood which we know to form so integral a part of the senile changes in the circulatory system. Tachycardia is thus not only, in many cases, an important sign of senile cardiac degeneration, but is also in itself an additional danger to the senile heart.

In infancy and childhood tachycardia is normal and physiological; in febrile diseases it is a never-failing symptom; in anæmia and other states of exhaustion, and in many diseases of the heart and blood-vessels, tachycardia is not an unusual symptom; while the other conditions with which this affection is found connected may be comprehended under two heads — intoxications and affections of the nervous system. *Conditions in which tachycardia may be present.*

The various intoxicants, or poisons, which give rise to tachycardia, comprise first of all — alcohol.

And in speaking of alcohol as a cause of tachycardia, no reference is meant to the ordinary rise of pulse that follows the use, or still more the temporary abuse, of alcohol, but solely to those cases of persistent rapid heart action, empty heart sounds, and feeble *Alcohol as a cause of tachycardia.*

pulse, which alone constitute the syndrome of tachycardia, and which are occasionally found in connection with chronic alcoholism.

In such cases sudden death not infrequently occurs, and the heart is found dilated and fibro-fattily degenerated. In these cases the tachycardia is believed to depend upon a neuritis of the vagus, due to the abuse of alcohol. Such cases are always serious, and are probably much more common than is as yet recognized. In some of them, and these the least serious, the heart and pulse are irregular as well as rapid, and the brain unaffected; in others the tachycardia either exists alone, or it may accompany delirium tremens, and it is then apt to be merged in what appears to be the more serious affection, while after all in the heart trouble the real danger lies, the nervous symptoms being of comparatively little consequence and quite appeasable by a twelve-hours sleep. As practitioners we are so apt to recognize a quick pulse as a usual accompaniment of the consumption of alcohol, and delirium tremens as a result that not uncommonly precedes the end, that we are apt to forget that the chronic abuse of alcohol originates a fibro-fatty degeneration of the myocardium, as well as a neuritis of the vagus, that the one impedes the cardiac function, and the other by paralyzing inhibition permits the heart to fly off in a hurry, impedes recovery, and duly recognized may be accepted as a measure of the danger present.

Tea and coffee used in moderation increase at first both the force and frequency of the heart's action, and induce a pleasant excitement of the cerebral functions, but the abuse of these stimulants produces in some actual intoxication, and in others that lowering of the blood pressure and acceleration of the heart's action which occasionally leads to an attack of tachycardia, during which the pulse is in some irregular. *Abuse of tea and coffee may be a cause of tachycardia.*

Tobacco is, however, that poison most largely abused by man, and from that abuse we gain a large experience. Nicotine, the poisonous alkaloid of tobacco, at first slightly slows the heart, or it may arrest it momentarily, causing intermission, or the inhibition may be strong enough to start the ventricle on its own independent rhythm, when irregularity soon follows (*vide antea*, p. 40). *Influence of tobacco in producing tachycardia.* When the dose is powerful enough to paralyze the vagus, the heart set free from its restraining influence starts off at a gallop, and we have an attack of paroxysmal tachycardia, with embryocardiac sounds, and increased precardiac dulness. The heart's action at times seems tumbling and irregular, but the pulse itself is small, feeble, and regular.

The following sphygmogram (Fig. 5) is an example of a hyperdicrotous, tachycardiac pulse of

Fig. 5.

low tension, beating perfectly regularly at the rate of 170 per minute, as reckoned by the sphygmograph.

Case of Tachycardia. This patient was suddenly seized with his tachycardia while playing a match at golf; he thought of giving it up, but a bumper of whiskey enabled him to win his match with what must have been a perfectly uncountable pulse, as even when at rest in bed this is never under 170 during an attack. This patient is now over sixty years of age, and during the last eight years he has had several similar seizures, all of them due to excessive smoking coupled with a good many nips of whiskey, the whiskey being never taken to excess. I have known his family for more than one generation, and not one of them has ever complained of the heart but himself, and he, indeed, resents his ailment rather than complains of it.

A rapid heart-beat means, as Donders first pointed out, a shortening of the systole,[1] a small amount of blood expelled by each ventricular contraction, hence shortening of the primary wave in the pulse-tracing and increased depth of the dicrotic notch, dicrotism of the pulse and pulse-tracing. When the pulse-rate is much increased, the pulse becomes hyperdicrotic; the ordinary dicrotic notch is carried on to the ascending limb of the tracing, and seems to be anacrotic, as in the sphygmogram here given. In this case there was a small, feeble, perfectly regular, but very rapid pulse (170), no dicrotism to be detected by the finger, but hyperdicrotism very evident in the tracing. There was increased precordial dulness, and a feeble, wobbling heart-beat, evidently a condition in itself not devoid of danger at any age, and one which indicates most unmistakably the risk to which a senile heart is exposed by an attack of tachycardia. Every dicrotic pulse is not a rapid one, neither is every rapid pulse dicrotic. But the amount of danger present in any case of tachycardia may be to a large extent measured by the degree of dicrotism present in the pulse, as this indicates diminution in the amount of blood expelled from the ventricle (contraction volume), increased residual accumulation, and tendency to

[1] *Nederl. Archiv. voor Genees-en Naturk.*, Bd. ii., 1865, S. 184.

death from failure of the heart — sudden or ingravescent asystole.

In sudden death from cardiac failure there is failure of the heart to contract, failure of the heart to respond to the call of the katabolic nerve — Asystole. At times, however, the failure to contract is not sudden and complete, but occupies an appreciable period of time, from a few moments to a few days, or even longer, and it is then most appropriately termed Ingravescent asystole.

Asystole may be sudden or ingravescent.

In neither of these forms of asystole is there any feeling of faintness — only a sensation of impending dissolution, and a gradual failure of both pulse and heart, the act of dying occupying but a few minutes, and the mind remaining clear to the last.[1]

When the asystole is of longer duration, the pulse is small, feeble, quick, and sometimes irregular; the heart's action is rapid, feeble, sometimes wobbly; the liver and spleen are congested, and if dying is prolonged, they may enlarge. There is œdema of both lungs, or œdema of one lung and effusion into the other pleura; there is often slight blood-spitting, from general pulmonary congestion or from local patches of pulmonary apoplexy due to thromboses; the œdema of the lung

[1] Balfour, *op. cit.*, p. 305.

is sometimes so great as to make the part affected seem solid, yet this solid œdema may disappear in a few hours, or it may shift its place when the position of the body is changed; there is slight œdema of the feet and ankles, with a slowly increasing soakage of all the tissues, a trace of albumen in the urine, which slowly increases, and a duskiness of the skin, which deepens as death approaches, and is most noticeable at the finger tips and nails. As a rule there is no recovery from this condition, though death may be lingering. This, however, depends of course upon the inducing cause; in olden times, when aconite was looked upon as the equivalent of digitalis, I have seen hearts brought into a state of almost fatal asystole by the one drug, quickly and rapidly restored to health by the other. Most usually this ingravescent asystole is a terminal phenomenon, and death long prepared for comes often unexpectedly at the last, the ingravescent asystole suddenly becomes complete.

Sundry medicinal agents also produce tachycardia when given in poisonous doses. *Digitalis*, for instance, when given in too large doses, or in doses too closely approximated, paralyzes the vagus and sets free from control the heart's idiomotor mechanism. If this paralysis comes on slowly, we have, first, a slow pulse with an occasional quick beat; by and by the pulse becomes quick with an occasional slow beat, or an intermission; and finally, when the regulating power is entirely lost, the intermissions disappear, and the pulse becomes regular but very rapid, the heart's sounds are embryocardiac, — reduced to a mere tic-tac, — the arterioles are dilated, and the blood pressure low.

Belladonna and *Atropine* in moderate doses increase the quickness, fulness, and force of the pulse; they also increase the blood pressure. In toxic doses both of these drugs paralyze the vagus, the heart runs off, and the pulse becomes extremely rapid, feeble, and often irregular.

How digitalis, belladonna, and atropine induce tachycardia.

Reflex tachycardia is generally of short duration, and is rarely attended by any danger. Reflex tachycardia is usually accompanied by other neurotic symptoms, such as dilatation of the pupils, flushing of the face, puffs of heat all over the body, or outbreaks of perspiration, local or general. These symptoms certainly indicate reflex action through the sympathetic system, but we must not, therefore, conclude that the tachycardia itself is produced by action on the accelerators alone, though there are certain other symptoms which seem also to point to this conclusion. For

Reflex tachycardia, its symptoms. How they are produced.

example, we find in reflex tachycardia that the heart's impulse is forcible and the pulse full, instead of both being feeble and the pulse small, as in ordinary tachycardia due to vagus inhibition. But we must not forget that the same thing also happens occasionally in tobacco poisoning, in which the heart hurry is undoubtedly due to vagus inhibition. In fact, remembering the trifling results, so far as tachycardia is concerned, which follow excitement of the accelerators alone, and also the fact that certain causes of reflex tachycardia do in other circumstances act as vagus inhibitors, the conclusion is forced upon us that in certain circumstances, not yet clearly understood, the same cause that inhibits the vagus also excites the augmentors, so that we have at one and the same time an idiomotor tachycardia from vagus inhibition, and a forcible heart-beat and a full radial pulse from excitation of the augmentors. In ordinary circumstances the vagus acts as the pendulum of a clock — it regulates the motion; when its action is inhibited, it is as if the pendulum were removed, and the idiomotor mechanism of the clock allowed to rattle on at an uncountable rate. This is tachycardia pure and simple. But *Difference between simple and reflex tachycardia.* when the pendulum is only shortened, not removed, the rate indeed is quickened, but considerable force of beat remains; this is reflex tachycardia.

There is nothing abnormal either of neurotic or organic origin which may not act as an excitant to an attack of tachycardia. Every kind of emotion or psychical impression; all sorts of neuroses, hysteria, epilepsy, neurasthenia; every form of dyspepsia; affections of the liver, only rarely; floating kidneys; prostatic disease; abdominal tumours, intestinal worms; various forms of neuralgia; also uterine affections of divers characters, —may all at times prove the exciting causes of paroxysms of heart hurry of shorter or longer duration, and in women these are most prone to occur during amenorrhœa and at the menopause. Affections of the lungs, and especially of the heart, are also well-known causes of tachycardia; and when the heart affection is a mitral stenosis, the heart hurry often persists for many years apparently without any serious detriment to the patient.

Whatever may be the exciting cause of the attack, there is no doubt that any breakdown in the general health, any anæmia that may be present, whether from increased hæmolysis or defective hæmogenesis, is a most powerful predisposing cause, especially if conjoined with that gouty venosity always present after middle life.

Tachycardia of purely emotional origin is often very persistent in its duration. In the case of a middle-aged lady in whom the attack was brought on by severe mental emotion of some duration culminating in a tragedy, it persisted for years, ultimately dying quite away. In this case the tachycardia was followed by a threatening of symmetrical gangrene of the finger-tips, which also was perfectly recovered from.[1]

Antecedent sources of emotion are common enough causes of tachycardia, but the connection is not always very obvious to the sufferer, and in all the complaints made it usually happens that the most important is never touched upon at all. In the case just referred to, the patient was almost well before the source of her sufferings was ascertained.

Two cases of emotional tachycardia.

The following case was of a similar character. This patient, a clergyman, consulted me several

---

[1] Raynaud's disease — another neurotic affection, of which this is the single instance out of many observed that showed any affinity to tachycardia. Symmetrical gangrene is more allied to those curious vaso-motor affections in which there is a persistent feeling of coldness, either local or general, which it is difficult to remove or even alleviate. The coldness seems to be due to actual constriction of the vessels. One of my patients died during the winter of what might be termed a universal chilblain. A feeling of local coldness, as well as pain, frequently precedes, generally accompanies, and is apt to follow even a threatening of vaso-motor gangrene.

years ago for rapid, irregular action, affecting a heart somewhat dilated and also hypertrophied. The trifling irregularity soon disappeared, and a spell of tachycardia set in that lasted, with some remissions, for a period of nearly four years. During all this time the pulse was continuously extremely rapid and feeble, the heart's impulse weak, and its sounds embryocardiac in character. With my finger on this feeble, rapid pulse, I have often felt it run off into a scarcely perceptible tremor,—*pulsus myurus*,—and while wondering whether it would ever return, it would suddenly come back with a feeble thump and continue on as before, the patient remarking, "That was one of my peculiar attacks," but never saying that he felt faint. Indeed, one of the most remarkable facts connected with this case was, that with a pulse so rapid—never under 130—and feeble, there was so little uneasiness or distress, and that the patient was able to go about very much as usual. Under appropriate treatment, coupled with several months' relief from duty, he was so far restored that at the end of two years he felt himself able to accept the most dignified position which his Church had in its power to bestow. And I may add that he discharged the somewhat onerous duties of this position not only with dignity and ability, but to the entire satisfaction of his friends and of his Church.

Nearly two years subsequently he died from an attack of pneumonia, the result of exposure to cold after exertion in early spring, having been wonderfully free from heart symptoms for some time previously. Indeed, the heart, being slightly hypertrophied, was not specially at fault at the last, though more than sixty years' service and all it had come through had not tended to improve its power of resistance.

The remarkable part of the case is this: that here we had a perfectly healthy man, leading a model life, and doing only the ordinary work of a country clergyman, which few would consider either hard or excessive, suddenly struck down with a serious attack of tachycardia engrafted upon a dilated and hypertrophied heart. There was an entire absence of all the usual causes of enlargement of the heart. There was no disease of the valves, no marked arterio-sclerosis, and therefore presumably no affection of the coronaries. There was no affection of the lungs or kidneys, nor had there been any undue exertion. The patient was well developed, and had reached advanced age in perfect health, so there was no reason to suspect abnormal narrowness of the aorta. Further, he was a most temperate man, so excess of any kind could not be alleged as a cause, and, so far as I could learn, there was no reason to suspect any hereditary tendency.[1] But he was over sixty, and the vascular changes which age brings on every one must have considerably progressed, when he was suddenly assailed by the most terrible bereavement which can befall any man. Then he began to age rapidly, and eighteen months subsequently he consulted me with the symptoms already described. Evidently the heart labouring, as all hearts do more or less under the strain thrown upon it by the loss of arterial elasticity, had its contractility impaired by the inhibitory emotional influence conveyed to it through the vagus. It must also at this time have suffered somewhat from impaired nutrition, and all these circumstances must have combined to produce the dilatation which was speedily followed by slight hypertrophy.

The heart hurry in this case did not die off in a few weeks or months, as is commonly the case in attacks of paroxysmal tachycardia, but persisted for years; and this we can scarcely wonder at when we remember that the cause was not only a powerful, but a persistent, emotion.

[1] *Vide* Traube, *Gesammelte Beiträge zur Pathologie und Physiologie,* Berlin, 1878-9; Strümpell, *Lehrbuch der Speciellen Pathologie und Therapie,* Leipzig, 1883, Erster Band, S. 422; and Oscar Fraentzel, *Die idiopathische Herzvergrösserungen,* Berlin, 1889.

This case is instructive as showing, in the first place, how efficient a cause of cardiac enlargement the mere natural loss of arterial elasticity is, even in those who are perfectly healthy and temperate. Just the other day I saw an old gentleman of eighty-two; in all his long life he had never ailed. He was of most temperate, almost abstemious, habits; up to a few months ago he thought nothing of walking five or six miles over a rough, hilly road, and was never breathless. I saw him for breathlessness due to pulmonary congestion following influenza, and, to my astonishment, found his heart dilated and hypertrophied, beating with a heaving, forcible impulse in the fifth interspace, considerably to the left of the nipple. As I had known this gentleman all my life, the condition of his heart was quite a revelation to me, and a very remarkable proof of the efficiency of natural causes in giving rise to cardiac enlargement, which in his case, even more than in most, seemed to deserve the adjective "idiopathic." The key to this case, as well as to all similar cases, lies in the structural change of the senile arteries, and in the fact that all such hearts are not simply hypertrophied, but are dilated and hypertrophied. Even Cohnheim has said that " the great majority of all idiopathic cardiac hypertrophies are eccentric," and that " non-eccentric hypertrophy has chiefly a theoretic interest," [1] — a statement that might be even more strongly emphasized.

In the second place, this case is interesting as showing how readily the erethism of a weak and labouring heart may pass into alarming, if not actually serious, tachycardia, under the influence of an overwhelming emotion.

And, lastly, this case furnishes a most remarkable example of the small amount of actual suffering entailed by even a most severe attack of tachycardia, and how wonderfully little the habits of life may sometimes be disturbed by what seems even to an expert to be a most serious cardiac affection.

To conclude, as vagus inhibition is the great cause of tachycardia, intra-thoracic tumours, often of no great size, pressing upon or involving the vagus in their structure, are well-known causes of persistent heart hurry, not simply paroxysmal, but fatal. .

[1] *Lectures on General Pathology*, New Sydenham Society's Translation, London, 1869, Vol. i., pp. 70, 71.

CHAPTER V

BRADYCARDIA AND DELIRIUM CORDIS

LAENNEC, the earliest of auscultators, has said, "We can distinguish two kinds of intermissions: the one *real*, consisting in an actual suspension of the heart's contractions; the other *false*, depending on contractions so feeble as to be imperceptible, or almost imperceptible, to the touch in the arteries."[1] And Hope has supplemented this by stating that "when one or two beats are regularly and permanently imperceptible in the pulse, such cases constitute the bulk of those in which the pulse is described by non-auscultators as being singularly slow — for instance, thirty or twenty per minute." And he adds, "In a few rare cases, however, it is really slow."[2] So far as my own experience goes the rarity has been all the other way, as I have seen many more really slow hearts, than hearts

---

[1] *A Treatise on the Diseases of the Chest, and on Mediate Auscultation*, translated by John Forbes, M.D., 2d edition, London, 1827, p. 570.

[2] *On Diseases of the Heart*, 3d edition, London, 1839, p. 377.

beating at the normal rate with an abnormally slow pulse, due to alternate hemi-systoles. There is never any difficulty in making a diagnosis between the two varieties of slow pulse; we have but to count heart and pulse together to realize that in the one class of cases each heart-beat, few and far between, is followed by a distinct pulse at the wrist, while in the other set a varying number of cardiac pulsations never reach the periphery. Sometimes every alternate beat is dropped, and at others two or more.

Slow pulse from hemisystole.

The first case of this kind that came before me was that of an old lady with a gouty history, but who had never had a regular attack, who was suddenly seized, while shopping, with what seemed to be an epileptic fit. In spite of what was supposed to be appropriate treatment, these seizures continued to recur whenever she made the slightest exertion, and when I saw her she was unable to rise from the recumbent position without bringing on an epileptiform attack. Upon examination, I found her pulse beating only 20 per minute, while her heart was beating at the rate of 60; only every third beat was strong enough to reach the periphery. The heart was dilated, with a feeble impulse, but without any murmur; the aortic second was

Case of false bradycardia.

accentuated. Remembering Stokes' admirable essay on the connection of pseudo-apoplectic attacks with the feeble circulation that he believed to depend upon fatty degeneration of the heart,[1] there was no difficulty in connecting the epileptiform seizures with the state of the heart, and just as little difficulty in determining upon the appropriate treatment. The result was most satisfactory — the old lady, who had been looked upon as the victim of serious senile epilepsy, had no more attacks. Within a week she was able to entertain some friends at dinner, and she lived for several years without any recurrence of her serious symptoms, dying gradually at last from asthenia.

Hearts, however, which are really slow belong to quite a different and a much more serious category. Several years ago I received the following letter from a professional friend: "A medical man in this neighbourhood, in extensive first-class practice, knowing that I had been your resident physician, asked me to examine his heart. What rather troubled him and made him think of his health was, that formerly his pulse was always 60, and that now it is invariably 48, except sometimes after dinner, *Case of true bradycardia.*

---

[1] *Dublin Quarterly Journal of Medical Science*, Vol. xi., 1846; also *Diseases of the Heart and Aorta*, Dublin, 1884, pp. 322, 362, etc.

if he has taken a little champagne, when it reaches 60 again. He has arcus senilis (age 53) well-marked, but nothing remarkable in the radial or temporal arteries, and is as active and energetic as possible. Lately I have noticed that he often looked tired and worn out, but he says he is not overworked. I carefully examined the heart, and found nothing except feeble apex-beat and sounds. His temperature does not, as a rule, come up to 98°. I tell you this, because Sir William Jenner told him that he had noticed that men with a slow pulse and rather low temperature live a long time. This gentleman does not feel at all ill, but is anxious to know whether his slow pulse (so much slower than formerly) ought to be looked upon as indicating degenerative changes in the heart and vessels; and if so, whether it would be wiser to knock off some of his work, which he can easily afford to do." My reply to this was, that the signs and symptoms detailed were evident indications of cardiac failure; that the heart, so far as my experience could enable me to judge without a personal interview, was beginning to dilate, that the arteries had undoubtedly lost their elasticity, and were probably even more atheromatous than was suspected. I advised considerable lessening of his daily work, and indicated the lines upon which the treatment should be conducted. This patient acted as advised; he survived for nearly nine years, and was then found dead in bed one morning when on a yachting tour.

A pulse of 48 is, of course, only abnormally slow in relation to the normal pulse of the individual, because, though 70 to 75 may be reckoned the normal pulse of most, there are some whose pulse never rises above 60, and a few — a very few — whose normal pulse is never even up to 48, and who yet enjoy perfect health. Haller tells of two people whose radial arteries did not beat oftener than from 24 to 30 times a minute; and M. Roux relates the case of an agriculturalist who had gone through his military service without difficulty, who never had a complaint either cardiac or cerebral, and who was a typical example of good health, and yet his pulse-rate was never over 34 to 40 per minute, and even a run of several minutes never raised it higher than from 50 to 55, and that only for a few seconds.[1] Several similar cases have been recorded, the most remarkable and best known being that of the great Napoleon, whose pulse, according to Corvisart, was only 40 per minute. Napoleon is often cited as an example of a slow pulse combined with perfect health; but Napoleon was an epileptic, like many

---

[1] *Vide* "Le pouls lent permanent." Par le docteur E. Leflaive. *Gazette des Hôpitaux*, 1891, p. 1072.

— if not most — of the sufferers from bradycardia.

Slow pulses are rarely to be found in early life, but occasionally they are found even at so early an age as five years; some of these youthful cases are apparently physiological and attended with perfect health, but the larger number at any age are strictly pathological, not only in their origin, but also and specially in their results. Rare at all ages, bradycardia increases in frequency and danger after middle life, and is more common among men than women. All the cases I have seen have been men.

Bradycardia may be physiological.

The earlier observers — Adams,[1] Richard Quain,[2] and Stokes[3] — endeavoured to connect sequentially a slow heart with fatty degeneration of the myocardium. Indeed, the sole survivor of these three still quotes slowness of the pulse as a symptom of this affection, acknowledging at the same time that quickening of the pulse increasing with age may also be an important indication of the same pathological condition.[4] But the very

But is most frequently pathological and senile.

---

[1] *Dublin Hospital Reports*, Vol. iv., 1827.
[2] *Medico-Chirurgical Transactions*, Vol. xxxiii., p. 102.
[3] *Op. cit.*, p. 326.
[4] *Dictionary of Medicine*, p. 595, 1882.

antagonism of the two symptoms precludes the idea of the connection of either with a fatty myocardium being anything but purely accidental. Indeed, a similar statement may be made in regard to atheromatous disease of the heart, aorta, or coronaries, as well as all other cardiac and vascular affections with which a slow pulse has been incidentally found connected. These lesions are all so much more frequently found apart from a slow pulse than with it, that it seems much more reasonable to conclude that the apparent connection is merely accidental, than that there is any direct relation of the one to the other. This is quite distinctly the case even in regard to the only lesion which is always present in every case of senile Bradycardia — dilatation and hypertrophy, the dilatation predominating. Slow pulses are rare, but after middle life dilatation and hypertrophy of the heart are of everyday occurrence.

Bradycardia not a result of fatty degeneration, nor of any other cardiac lesion.

Inhibitory impulses, we know, pass through the inhibitory centre down the vagus to the heart; these slow the heart and diminish its excitability. Roy and Adami tell us that there is a limit to this slowing, and that after a longer or shorter period the ventricles start off on an independent rhythm of their own (*vide antea*, p. 39). Accident, however, frequently carries out experiments which are more suggestive and often more fruitful than any contrived by art, and this seems to be specially true in relation to the causation of slow pulse.

Surgical observers have long since recognized that fracture of the cervical vertebræ, especially of the fifth or sixth, frequently gives rise to slow pulse. Gurlt says that fractures even as low down as the first dorsal vertebra may have this result, and that the pulse may fall as low as 36 or even 20 per minute;[1] and Charcot states that retardation of the pulse is one of the most interesting and least noticed facts of the symptomatology of cervical spinal lesion.[2]

Relation of injury of the cervical cord to slow pulse.

Jonathan Hutchinson tells us that unless injury to the spine is in the cervical region, no influence on the heart's action is ever observed. But he states that if the fracture is high up, the cardiac pulsations are greatly diminished in frequency, while (from the paralysis of the artery) the pulse itself is remarkably full and large. He adds that it is very remarkable to see a man screaming with pain and obviously suffering acutely,

---

[1] *Handbuch der Lehre von den Knochenbrüchen*, 1864.

[2] *Lectures on the Diseases of the Nervous System*, by J. M. Charcot, New Sydenham Society's Translation, London, 1881, p. 117.

with a full, slow pulse, beating not over 48 per minute.[1]

Rosenthal has recorded the case of a girl of fifteen who received a blow on the region of the sixth cervical vertebra. This was followed by symptoms of slight and quite transitory cerebral shock, accompanied by hemiplegia of the right side, which did not last longer than twenty-four hours. But for four weeks subsequent to the injury, the pupil (presumably the right, but which is not stated) remained dilated, and the cardiac pulsations oscillated between 56 and 48. The patient recovered completely.[2]

This fact of slow pulse following injury to the cervical cord, and passing off when that injury is recovered from, may, I think, be very instructively considered in connection with Holberton's well-known case, in which the injury to the cervical cord was not direct, but the result of inflammatory action, and in which it took two years to develop retardation of the pulse.

This gentleman, aged sixty-four, was thrown on his head in the hunting-field in December, 1834. At first he was stiff and sore, with great pain in the neck, about the cuneiform process and the condyles of the *os occipitis*. The pain continued

---

[1] *London Hospital Reports*, 1866, p. 366.
[2] Charcot, *op. cit.*, p. 117.

about six weeks. At the end of a year, he was well, in excellent spirits, but still complaining of a difficulty in moving his head.

In January, 1837, he had a fainting fit when out walking, and the medical man who attended found his pulse to be only 20 in the minute. His usual pulse was now found to be 33, but often during a fit it fell to 20, 15, or 8 in the minute, and even when not in a fit, it was occasionally as low as $7\frac{1}{2}$ per minute. His syncopal attacks always ended in epileptiform seizures, and as time went on, they increased in frequency as well as in severity. His first alarming succession of fits occurred in June, 1838, and his last and fatal attack was in April, 1840. After death his heart was found to be enlarged, the walls of the left ventricle were rather thin, the valves healthy, the auriculo-ventricular opening dilated. No ossification or calcareous deposit was found in any part of the vascular system. The inflammatory action which had followed the injury to the first and second vertebræ had narrowed the *foramen magnum* and upper part of the spinal canal, compressing and increasing the density of the *medulla oblongata* and upper part of the spinal cord. This gentleman never had any paralysis, never after the first few weeks suffered pain in the neck. His spirits when free from attacks were excellent, and his general health often very good. During the last three or four years of his life he was liable to cold feet, and suffered from a feeling of general chilliness.[1]

These cases which so markedly connect slight and transient injury (concussion) of the cervical cord with temporary slowness of the pulse, and more serious and permanent injury of the same part of the cord with permanent slowness of the pulse, leave no room for doubt that through this centre it is possible to convey to the heart an inhibitory influence, powerful enough to bring its pulsations down to $7\frac{1}{2}$ beats per minute, and persistent enough to last for many years.

Roy and Adami tell us that vagus inhibition may arrest ventricular action altogether for a short period, but that it does not persistently slow the heart, because sooner or later a time arrives when vagus inhibition is set at naught, and the ventricles start off on a rhythm of their own, an idio-ventricular rhythm[2] (*vide antea*, p. 39). But cervical inhibition, as we may call it, is not only strong enough to force a slow rhythm upon the heart, but is also powerful enough to compel the heart to

Difference between vagus and cervical inhibition.

[1] "A case of slow pulse with fainting fits, which first came on two years after an injury to the neck from a fall." By T. H. Holberton, *Medico-Chirurgical Transactions*, London, 1841, p. 76.

[2] *Op. cit.*, p. 233.

keep to this slow rhythm for years, with but trifling variations. For years the heart may pulsate at the rate of 20, 30, or 40 beats per minute, without ever quickening its pace, without an intermission, or even a hint at irregularity. It seems as if the whole heart, sinus, auricle, and ventricle, were forcibly controlled and compelled to keep steadily to the unnatural rhythm. Now and then, as in Holberton's case, we have an occasional intermission. Still more rarely we have a bout of irregularity interposed, as in a most interesting case which I shall presently relate. But as a rule, the steady, slow, funereal beat never varies from the time it commences till the patient's death.

If we ask why the cervical cord should have so potent an influence upon the heart, there seems to be but one possible answer: Because from this region the *spinal accessory* arises. This nerve rises by several roots, beginning as low down as the sixth cervical vertebra; it runs up within the spinal canal through the *foramen magnum* into the cranial cavity, and thence it passes out through the *foramen lacerum posterius* in close proximity to the vagus. The internal portion of the spinal accessory subsequently joins the vagus and is distributed to the heart, presumably as its motor nerve. The vagus

Region of cervical cardiac inhibition is that from which the spinal accessory arises.

and the *nervus accessorius* resemble a spinal nerve, the vagus with its ganglion being the posterior or sensitive root, while the spinal accessory is the anterior or motor root. Concussion of the cord at the origin of the spinal accessory produces temporary slowness of the pulse; severe injury to that part of the cord, disease of the cervical membranes, or of those at the base of the brain (pachymeningitis), involving injury or compression of the accessory nerve, produces permanent slowness of the pulse. Besides these direct injuries there are various reflexes and several poisons which are supposed to have a retarding influence upon the heart. The Indian fakeers, it is alleged, slow the heart and diminish the force of its beat by voluntary compression of the muscular branches of the *nervus accessorius* in the neck (*vide antea*, p. 67). Various cases of slow pulse have been recorded in connection with abscess of the brain; gastric irritation and constipation often precipitate the syncopal attacks, and by some have been supposed to be the only exciting cause. In one of my own cases alcoholic excess was the only possible cause that could be discovered. Holberton's patient had his first serious attack the day following a heavy dinner, when, as Holberton says, he "had eaten heartily of a variety of substances," and

Gastric irritation and constipation great provocatives of syncopal attacks.

with him both gastric irritation and constipation were found to be serious provocatives of syncopal attacks, and they always affected the pulse-rate, either raising it or lowering it, and, strange to say, the one was as liable as the other to be followed by an attack. Burnett also records a case of slow pulse with epileptiform seizures, in which the only discoverable cause was disturbance of the chylopoietic viscera; and he quotes two similar cases from Morgagni, in which no other cause could be discovered. In Burnett's own case the pulse ranged from 14 to 28, though it occasionally rose to 56.[1]

Several of my patients have died in syncopal attacks, but I myself have never seen such a seizure. Holberton describes a fit as always preceded by cessation of the pulse for a second or two before syncope took place; on the heart recommencing to beat, "the face would redden, and consciousness return with a wild stare and occasionally a snorting, a slight foaming at the mouth, and a convulsive action of the muscles of the mouth and face."[2]

Character of a syncopal attack.

The initiatory seizure seems thus to be essen-

---

[1] "Cases of Epilepsy attended with Remarkable Slowness of the Pulse," by William Burnett, M.D., *Medico-chirurgical Transactions*, 1827, p. 202.   [2] *Loc. cit.*, p. 79.

tially syncopal in character, while the succeeding phenomena are evidently due to the unusually large blood-wave with which the tissues are suddenly flushed on what may be termed the return of life.

But however effectual affections of the chylopoietic viscera may be in the production of syncopal attacks when a slow pulse already exists, the numbers of disturbed stomachs and constipated bowels that are found apart from any retardation of the pulse, make it extremely doubtful — to say the least of it — whether of themselves these conditions have any material effect in slowing the pulse. And the same remark may be made in regard to all those reflexes to which retardation of the pulse has been assigned as a symptom. More definite information as to this is still a desideratum. *Reflex heart retardation doubtful.*

We know that many poisons, both organic and inorganic, bile, uræmia, diphtheria, digitalis, lead, etc., slow the heart, but these all have a direct action upon the nerves, and upon the nerve-centres.[1] Indeed, all the information at present at our command seems to point to direct action on the spinal accessory in the neck or *Retardation of the pulse probably always due to direct action on the nervus accessorius.*

---

[1] Greenhow mentions a remarkable case of slow pulse with paralysis following diphtheria, in which the large nerves of the

chest, before or after its junction with the vagus, whether by concussion, compression, or otherwise, as undeniably the most potent, and probably the only cause of abnormal or pathological bradycardia.

Hemi-systole has already been mentioned as a cause of apparent slowness of the pulse, because only every second or third beat is strong enough to reach the periphery. The following sphygmogram represents this condition. In it (Fig. 6)

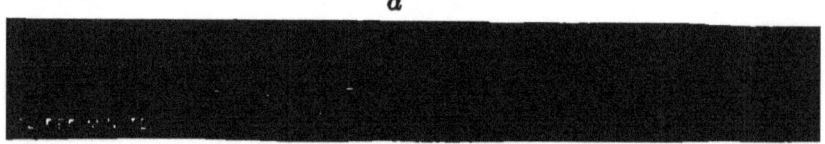

Fig. 6.

the pulse is seen to rise at once to its full height; the secondary dicrotic wave occupies its usual position, but lower down the descending limb — just where in a normal tracing the elevation of a new pulse ought to begin — there is a slight elevation (*a*) due to the hemi-systole, imperceptible to touch and not always to be found in the tracing.

In true bradycardia the sphygmogram is perfectly different. In it (Fig. 7) there is what appears to be a great round-topped predicrotic

---

limbs were painful to touch; the natural inference is that probably the spinal accessory was similarly affected. Recovery was complete. — *Lancet*, 1872, Vol. i., p. 615.

blood-wave, as if the blood pressure was greatly increased, or as if the rigid arterial wall was only slowly raised by the advancing blood-wave.

Fig. 7.

The true explanation is as follows: So long as the circulation remains intact, the heart gets more distended the longer the diastole is prolonged. At each systole a larger blood-wave than usual is thrown out, and as the arteries have *Explanation of a bradycardiac sphymogram.* had a longer time than usual to empty themselves, it passes rapidly onwards, and as can be readily understood, the secondary dicrotic wave is not only of greater amplitude than usual, but it also occurs earlier on the descending limb. In the sphygmogram (Fig. 7) the point A marks the height of the pulse-wave. The round top following is not, as might be supposed, the pulse-wave itself, but is really the secondary or dicrotic wave placed near the upper part of the descending limb instead of about its middle. In some sphygmograms, this dicrotic wave is so ample and so premature that it appears to occupy the very summit of the wave, the true apex of the pulse-wave lying below it, so that the tracing has an anacrotic appearance.

The large blood-wave sent on is naturally associated with a temporary rise of blood pressure, which rapidly dies off through the continuous outflow through the arterioles during the prolonged diastole. Hence in bradycardia we have, as in aortic regurgitation, an abnormally high blood pressure alternating with an exceptionally low one. A knowledge of this explains much that seems anomalous in the history of bradycardia, and it has also a not unimportant bearing on the treatment of such cases.

In the sphygmogram (Fig. 7), the pulse-rate was 32, but it varied from 36 to 28, and in this patient, as in all the senile bradycardiac hearts I have ever seen, there was marked dilatation of the heart, extension of the precordial dulness, apex-beat to the left of its usual position, and always a mitral murmur — generally a systolic murmur — in all the areas.

Senile bradycardia probably always associated with cardiac dilatation.

Knowing as we do the very high blood pressure the heart has to cope with shortly after the commencement of systole, the fact that most of these slow hearts belong to the latter half of life, and that the heart, in common with the other tissues suffers in its nutrition from the extremely low blood pressure prevailing during diastole, and suffers most just when it is called upon to make its greatest exertion, we cannot wonder that such hearts are always dilated. They also hypertrophy — never much, but a little — quite sufficient to enable them to carry on the circulation. I have never seen any reason to regard the myocardium of these slow hearts as specially feeble, — rather the reverse. But sufferers from senile bradycardia are generally sluggish and inert, which is perhaps not to be wondered at.

About a dozen years ago I received the following letter: "In autumn, 1875, after a time of much anxiety, I fell down and was unconscious for two minutes, with a very slow pulse. At various times after that, in 1877 and in 1878, I had turns of faintness, accompanied by great slowing of the pulse, which resumed its natural pace when the faintness wore off. In November, 1879, the pulse got down to a steady slowness of 36 per minute. A course of quinine and iron was tried without any good effect. My friend, Dr. Dobie, of Chester, then prescribed for me, and after about six weeks, about the middle of February, the pulse was suddenly restored from 36 to 70, and continued at its usual rate all March, but in April it fell gradually back to 36, keeping remarkably steady at that figure. Occasionally, for a few minutes at a time, it rose to 40

Case of bradycardia.

or fell to 28, but it speedily returned to 36 as its normal rate, which it has ever since maintained. During June and July, I again tried Dr. Dobie's prescription, but without any good effect. In August I went to Harrogate, and by Dr. Myrtle's advice took Kissingen water. At the end of a week he supplemented this with chloride of iron water. At the end of another week I had become rapidly weak, and Dr. Myrtle ordered me to abandon this prescription. Since then I have abandoned all treatment, and continue very weak. This day, at early morning, my pulse was 30; while I write it is 36."

This letter was speedily followed by a personal visit from the patient himself, and I find, from notes taken at the time, that he had a weak dilated heart, with a loud systolic mitral and tricuspid murmur; pulse ranging from 36 to 40; no albuminuria. He made but little progress while under observation, the pulse still continuing slow, and he was lost sight of in a few months. Being lately — 1890 — desirous of ascertaining the result, I put myself in communication with the patient's friends, and received from himself the following letter:

"I seem to have sent you an account of my illness in 1880, so I need not notice it previous to that date.

"During the years 1881, '82, and '83, the pulse

continued from 30 to 34, accompanied by great exhaustion. During these years I gave up all treatment of any kind, living in my usual way, without any medical advice whatever.

"About the end of November, 1883, I was amusing myself with a little grandson from India, and had a good deal of laughing and fun with him. A change seemed to have come upon the long dreary tramp of 30, with its solemn regularity; it had now become of the most irregular character, ranging from 30 to 80. A strong beat, then five or six very small ones all in a rabble, like the bursting of a wooden barrier across a river, with masses of the debris gathering again and obstructing the current for a time, and then bursting through again. At the end of about a week the irregularity ceased, and at the end of 1883 it was moving quite naturally at 70.

"This improved state of things continued for several months, when it began again to slow down to the thirties, where it has continued ever since; the highest record of this period being about 36, and the lowest, 28. One peculiar feature of this slow pulse is its regularity; in some conditions 36 can be depended upon, in some 34, and in some 32. It does not often come below this, although 28 has been recorded several times. While I write my pulse is perfectly steady at 36.

I have been much troubled with sleeplessness, caused by twitching or flickering of the legs, accompanied by great depression of spirits; but in another three months I shall, if spared, have lived the threescore and ten, which most people admit is quite long enough. I always looked well, having a very florid complexion, with a good deal of blue, however, in it. I have not fallen down again to be insensible as I was at the first, but have been glad to stretch myself out on the road sometimes, so as to avoid what appeared to be a fainting turn coming on."

This patient died two years subsequently in a syncopal attack, his heart having been irregular for some time previously, in this way: that for one while the beats were very slow, and again, for another while, faster, but always slow. There was no examination of the body.

I am not aware of any other case of true brady-cardia in which marked irregularity was even an occasional phenomenon, but to my knowledge there is no other recorded experience of the effect of unwonted exertion on a heart of this character.

The effect so graphically described seems to have been the result of the unwonted exertion forcing the ventricles into an independent rhythm. When the systole of that independent ventricular rhythm happened to coincide with the systole of the auricle, then there was the occasional "strong beat" referred to; while the "rabble" of five or six small beats were the result of ventricular systoles which did not coincide with any auricular systole.

While *delirium cordis* of this character is an unusual symptom in true bradycardia, it is by no means uncommon in gouty, dilated hearts, at least as a temporary phenomenon. It is always desirable in all such cases to make a careful comparison between heart and pulse, and if possible to take a sphygmogram of the latter. *Delirium cordis* is common enough in mitral stenosis, but in that affection it is never so striking a phenomenon as in the dilated gouty heart. Because in stenosis — unless the stenosis is very slight — the difference between the size of the beats is never so marked as when the auriculo-ventricular opening is at least of the normal size. Less frequently this *delirium cordis* in the gouty heart is found to be constantly present, never ceasing from its first appearance till death occurs. Of my own personal knowledge I can only recall three such cases, all of them well-marked examples. Two of these were well-known professional men, who both died from dilated hearts — one at the age of threescore and ten, and the other twenty years younger.

Delirium cordis.

The elder of these had the pulse of *delirium cordis* for twenty years before his death, of my own knowledge. I doubt if the younger man suffered for longer than about five years, and both continued to work till close upon the end with the utmost calmness and self-possession. The only other example of well-marked and persistent *delirium cordis* I can now recall was an old lady shown to me as a clinical curiosity, who lived in the heart of Westmoreland, and did not seem in the least disturbed by her unusual condition, of which she was yet fully conscious.

CHAPTER VI

ANGINA PECTORIS

ACCORDING to Quain 80 per cent of all cases of angina pectoris occur after the fortieth year of life; there can be, therefore, no hesitation in regarding it as a symptom of the senile heart. Yet even when childhood is scarcely passed life may be cut short with this symptom, and it is equally certain that death may then occur from causes usually regarded as purely senile in character. Tortuous, hard, and atheromatous arteries are not uncommon in early life, and Dr. Wild of Manchester has recorded the sudden death of a girl of twelve with advanced sclerosis of the coronary arteries.[1] She was not known to have suffered from angina, but there is such a thing as *angina sine dolore*, and sudden death with such a lesion, and without other evident cause, may very fairly be attributed to this.

Angina pectoris a symptom of the senile heart.

Yet death from angina may occur when childhood is scarcely passed.

[1] *The Manchester Medical Chronicle*, July, 1892, p. 230.

Wild has also recorded the sudden death of a girl of nineteen from angina,[1] and I myself have published a case of death from angina at the early age of twenty-four.[2] No age can therefore be looked upon as necessarily free from lesions usually found in advanced life, nor is any period of life always exempt from symptoms commonly found in connection with senile lesions.

The term "pseudo-angina" is often applied to anginous pains occurring before middle life, especially in the female sex, and yet we see that fatal angina may occur in one who is still but a girl. To talk of pseudo-angina is, however, a mark of ignorance rather than of refinement of diagnosis; for angina is but a symptom, and if well-marked, it should no more be stigmatized as "pseudo," because it occurs in youth, than the lesion with which it is sometimes associated should be called functional because it happens to be curable. At the same time there are plenty of pains to be found about the left side of the chest, and even in connection with the heart itself, which are not angina, and these it is of importance to differentiate for the patient's comfort as well as for his treatment.

Angina a symptom which may occur at any age.

Many precordial pains not anginous.

[1] *Op. cit.*, May, 1889, p. 146. [2] Balfour, *op. cit.*, p. 300.

Constipation dependent on torpor of the colon, especially if associated with chlorosis, is not infrequently accompanied with neuralgic pains radiating from the neighbourhood of the *scrobiculus cordis* over the edge of the false ribs, and sometimes shooting into the cardiac area itself. The pain in such a case is constant with occasional exacerbations; it always radiates from some part of the colon, and may shoot round the chest, or even into the cardiac area, but never upwards or into either arm; it is not increased by exercise, nor does it get worse during night. The heart may have all the usual chlorotic murmurs, but the pulse is always soft and compressible. This neuralgia is curable, but not always readily so.

Torpor and congestion of the liver, so constant an accompaniment of gastro-duodenal dyspepsia, is often associated with pain below either clavicle, about the second interspace. This probably arises from irritation of the phrenic nerve shooting as pain into the upper intercostal nerves. On the right side this simulates lung disease; on the left side it is apt to be mistaken for a heart pain.

Intercostal myalgia and neuralgia often encroach upon the cardiac area and get referred to the heart; also acute commencing pleurisy, often free from friction because movement is so painful, if near the cardiac area, gets talked of as a heart pain, though here the thermometer helps to keep the diagnosis right. The heart itself often suffers from burning, stinging, or cutting pains, the exact nature of which it is not always easy to determine, but which probably are always either of a rheumatic or gouty character — most probably gouty. Finally, there is a cardiac pain dependent upon pressure on the cardiac nerves. If the tumour, whatever its character, which produces this pressure and pain is too small to be detectable, and especially if it occurs in youth or early middle life, the pain itself is apt to be stigmatized as a spurious angina, as a mere neurotic pain to be fought against. And yet the ailment may be serious enough to cause sudden death erelong, and of this I have seen several instances. It is easy enough to separate a pain of this kind from true angina, if we get the chance of seeing a paroxysm, but then the difficulty begins. In true angina the danger is great, and the prognosis always serious, because true angina depends upon an interference with the function of the katabolic nerve, and in its mildest form instantly threatens the citadel of life. But in what we may term — for want of a better expression — false angina, we have only to deal with pain, the danger of which depends upon its cause; if the pain is caused by the pressure of a gland, the danger may be but slight; but if it be caused by a small substernal aneurism, the danger is great and imminent. In a few such cases it is possible to make a fairly accurate diagnosis; in others, this is absolutely impossible. There is no class of cases in which greater care and circumspection are required, and even with the largest experience an error may be committed; for the patient's sake it is better to err in excess of caution. Various authors — amongst them, Anstie[1] and Huchard[2] — have laid down certain rules for simplifying the diagnosis between true and false angina, Huchard especially has entered very fully into the question; but there is not one of the many indications commented upon which is not liable to serious exception, and with the greatest care doubtful cases will always occur in which a perfectly accurate diagnosis seems impossible. In saying this I refer specially to one case well known to myself as well as to others. In the case referred to there is no suspicion of hysterical exaggeration of any of the ordinary neuralgiæ, described as occasionally implicating the region of the heart. This patient was formerly a nurse in one of the largest hospitals in Britain, and has

---

[1] *Neuralgia and its Counterfeits*, London, Macmillan & Co., 1871, p. 75.

[2] *Maladies du Cœur et des Vaisseaux*, Paris, 1893, 2me ed., p. 719. Huchard has gone very fully into the whole subject.

been seen in her attacks by some of the ablest physicians, who have always treated her as suffering from true angina. She has long been happily married, though without family, and marriage has neither increased nor diminished the frequency or intensity of her attacks. During the last twelve years I have repeatedly seen and examined her, and I have always doubted the reality of her seizures.

Case of doubtful angina.

She has pain, no doubt, but, though a woman over fifty, she is too healthy and blooming to suffer from true angina. I have, however, never had an opportunity of seeing one of her attacks. Only to-day — as I was writing this — she called and told me that of late she had been suffering from gouty pains in her joints, especially in her fingers and wrists; and she added, "While I have these pains I am so irritable that I am a nuisance to myself and to every one about me; and what puzzles my doctor as well as myself is that while I have these pains I have none of my old attacks, but the moment the pains leave my joints I get one of my old attacks." Obviously this is — after all these years — the clue to her case; evidently she has a recurrent gouty neuralgia of the heart, an angina, no doubt, of a kind, yet neither a true nor, properly speaking, a pseudo-angina.

As a rule it is not difficult to differentiate all the varieties of cardiac pain from angina as well as from one another, though, in a doubtful case, the observation of an attack — when that is possible — may be of the greatest assistance.

We must not forget that in the syndrome of angina pain — even though severe — plays but a subordinate part, while in all those other affections which simulate it pain is the prominent and paramount symptom.

In a severe attack of angina, the patient dare scarcely breathe till the pain abates, not because of the pain, but by reason of that awful sense of impending dissolution of which the pain is, as it were, the subjective symbol. But all attacks are not so severe, and a certain amount of jactitation is sometimes observed, while in the angina associated with aortic regurgitation, forced inspiration and violent movements of the arms are occasionally resorted to with the object of relieving the agonizing pain, and sometimes successfully.[1]

As some of the acknowledged causes of angina may be present in the young as well as in the old, we are justified in regarding as true angina any paroxysmally recurring cardiac pain which cannot be referred to any of the varieties of neuralgia

Case of true angina in a young woman of twenty-five.

[1] Vide Balfour, *op. cit.* second edition, 1882, pp. 273 and 306.

just described, even although it occurs in a young person, and may be associated with more or less jactitation. Of this there could scarcely be a better example than the following case: In September 1888, a young married woman was sent for my opinion, with the following history: "M. S., æt. 25; married five years ago; has had two children, the last of them a month ago; has hereditary predisposition to angina. About eight years ago she had diphtheria with pericarditis, from which she made a good recovery under the care of the late Dr. Kelburne King. She was married five years ago, became pregnant, and during the first six months she suffered much from attacks of syncope. She made a fair recovery from childbirth. During the last three and a half years she has suffered increasingly from syncopal attacks, preceded by or accompanied with pains of an anginal character. She derives considerable benefit from nitrite of amyl inhalations, which cut the attack short. She was confined of her second child about a month ago, and recovered strength very slowly, until digitaline was administered, when she improved rapidly." I found Mrs. S.'s heart well contracted and slightly thumping in its action from three of Nativelle's granules having been taken daily for some time. She stated that her anginal attacks were always preceded by pallor of the face and fingers, that she could move about freely during the attack, and that it was always relieved by stimulants or by nitrite of amyl inhalations. The symptoms and history of this case showed it to be one in which the attacks were probably due to arterial spasm raising the blood pressure, and thus throwing an undue strain upon a feeble spanæmic heart which had been somewhat dilated; the nervous phenomena being obviously due to the instability of the nervous system, an instability the result of imperfect nutrition. Evidently a case liable to be branded as a hysterical or pseudo-angina, but really a case of true angina occurring in a young neurotic female, from a curable cause, and with, therefore, a favourable prognosis. Two years subsequently I enquired as to her progress, and received the following reply: "I may report favourably as to Mrs. S. The anginous character of her attacks gradually became less marked, and her general health much improved. She has since had another baby, making a good recovery."

I have called this a case of true angina, and such it undoubtedly was, meaning by angina a pain of the heart induced by a call for increased exertion, as in this case from a reflex rise of blood pressure, as in others from mere bodily exertion.

With this conception of angina, we can understand that it may vary much in degree. Probably the slightest possible form of it is the sharp pain that occasionally accompanies the augmentor action following an intermission, or a short spell of *tremor cordis* in a spanæmic heart.

Pain of angina may vary much.

Apart from the trifling form just referred to, the pain of angina varies from a dull agonizing ache, to a feeling as if a mailed hand grasped the chest in the cardiac area and squirted through its fingers flashes of excruciating agony up to the left shoulder joint, sometimes into both shoulder joints, extending down to the elbow or along the ulnar nerve to the fourth and third fingers on the left or on both sides. Occasionally the pain shoots up the neck, generally on the left side; or into the *scrobiculus cordis;* more rarely it shoots down the loins and legs. The sufferer has a feeling of choking, but the breathing is perfectly free, and is only restrained by the dread lest the slightest movement should precipitate the end which seems so terribly near. The countenance may be pinched, ghastly pale, and covered with beads of perspiration (*facies Hippocratica*). But often the face is quite unchanged, save only for an anxious, haggard expression. In the angina that occasionally complicates aortic regurgitation or indicates substernal aneurism, as well as in that associated with other more curable cardiac affections, the pain is more acute, less oppressive and appalling, and it is sometimes conjoined with so much jactitation as to simulate a pure neurosis.

The causes of angina may seem to be various, but they are all of a kind to depress the dynamic force of the nerve implicated, or of the heart itself, which is the automatic source of its own energy. Pressure on some of the nerves of the cardiac or aortic plexus is not an infrequent cause of angina. This pressure may be produced by a tumour, often a very small one; by a small substernal aneurism, which usually escapes detection; or by a dilated aorta, sometimes without, but more commonly associated with, regurgitation through the semi-lunar valves. *The cause of angina always some depressant of nervous or cardiac energy.*

One of the most common concomitants of angina is sclerosis of the coronary arteries; indeed, so common is the conjunction that the arterial sclerosis has often been looked upon as the cause of the angina. But coronary sclerosis is too often present, where there never has been any angina, to permit the concurrence being looked upon as anything more than accidental. Fatty degeneration of the myocardium is often found where angina has been present during life, and it too has been supposed to be a cause of the angina;

but, like arterial sclerosis, fatty degeneration of the myocardium is very often found where there has been no antecedent angina. Like coronary sclerosis itself, therefore, fatty degeneration of the myocardium can only be regarded as a concomitant of angina, and not as a cause. On the other hand, fatty degeneration of the myocardium is due to faulty metabolism from an imperfect blood-supply to the part affected; an imperfect blood-supply is a common result of arterial sclerosis, and is in fact the connecting link between coronary sclerosis and fatty myocardium. It is the one common factor these two conditions have — of the one it is a result, and of the other a cause. And when we inquire into the matter, we find that an imperfect blood-supply is a factor common, not only to the conditions just referred to, but also to almost every condition of heart with which angina has ever been found associated. Among these we may reckon embolism and thrombosis of the coronary arteries — diminution of the calibre of these vessels at their origin at the root of the aorta, or in their course through the heart, by inflammatory, atheromatous, or syphilitic processes. On rare occasions the heart in late life becomes enlarged beyond the feeding powers of coronary arteries congenitally deficient in size or in number, as happened

Angina almost invariably the result of cardiac ischæmia.

in the case of Dr. Arnold.¹ More commonly, simple dilatation and hypertrophy get in excess of the feeding powers of the ordinary coronaries, because of some failure in the quality of the nutriment supplied.

A good deal has been said about tobacco and tea as causes of angina, especially by French writers, as if the nicotine and theine produced it of themselves by a special act of poisoning. But angina from these causes, as well as from gastric derangements, is a rare accident, and never happens unless there has been some previous spanæmia, and some slight dilatation of the heart. It accompanies, or rather follows, some preceding irregularity of the heart's action; this we know involves

Tobacco, tea, and gastric derangements produce angina only by enfeebling the heart, and inducing relative ischæmia.

---

¹ The size of the coronaries is quite disproportionate to the mass of muscle to be fed, so that the heart may be looked on as having an excessive supply of blood compared with other muscles.— Odriozola, *Étude sur le Cœur senile*, Paris, 1888, p. 5. Dr. Arnold died at forty-seven of his first attack. "The heart was rather large. . . . The muscular structure of the heart was in every part remarkably thin, soft, and loose in its texture. The walls of the right ventricle were specially thin, in some parts not much thicker than the aorta. . . . Its cavity was large. The walls of the left ventricle, too, were much thinner and softer than natural, and the muscular fibres of the heart generally were pale and brown. . . . There was but one coronary artery, and, considering the size of the heart, it appeared to be of small dimensions."— Latham, *Diseases of the Heart*, London, 1846, Vol. ii., p. 377.

lessening of the ventricular output, with consequent residual accumulation and ventricular dilatation (*vide antea*, p. 41). To empty the ventricle in this condition the augmentor nerve is called into play, and this call for extra exertion is the incitation to angina. Not because tobacco and tea are poisons specially incentive to angina, but because their abuse has so lowered the health and impoverished the blood as to enfeeble the myocardium and induce a relative ischæmia, an ischæmia of quality though not of quantity. Spanæmic blood involves imperfect nutrition, and as the energy of the heart depends upon the perfection of its metabolism, long continuance of imperfect nutrition implies a commensurate loss of cardiac energy.[1] After middle life this is always a serious matter, and even in youth it is not devoid of danger, and may precipitate a fatal issue.

Danger of imperfect cardiac nutrition.

The vigour of a muscle may vary from nothing to a maximum, and depends upon the perfection of its metabolism. An ordinary skeletal muscle only possesses irritability towards stimuli; but the heart has not only the power of originating spontaneous rhythmic movements, but is also able to

---

[1] Von Bezold, *Untersuchungen aus dem physiologischen Laboratorium in Wurzburg*, Leipzig, 1867; Erster Theil, S. 279, etc.

store a reserve of energy so great that, in some animals, these spontaneous movements go on for hours after the heart has been separated from the body. It is evident, therefore, that the metabolism of the heart is of a very much higher order than that of the skeletal muscles, and is all the more readily affected injuriously by any changes in the quantity or quality of the blood which furnishes its basis.[1] The large reserve of energy with which the heart starts on its extra-uterine life, and which is always maintained during healthy life, enables it at any age long to resist hurtful influences of this character, but in time they tell.

When a bad bout of irregularity or intermission, induced by mental emotion or any other cause, or when such an increase of muscular exertion as is involved in going up a stair, or any acclivity, or when any sudden rise of blood pressure, from reflex causes, calls for increased action in a heart with its energy impaired by malnutrition from long-continued spanæmia, by positive obstruction to the coronary circulation, or, as is more frequently the case, by a combination of both, the response may be so imperfect that the function of the augmentor nerve is sensibly impeded. The call for increased katabolic action is at once followed by sudden

Cause of a fit of angina.

---

[1] Foster, *op. cit.*, p. 344.

exhaustion, and this is revealed as an agonizing pain beneath the sternum, that shoots along some or all of those sensitive spinal nerves with which the sympathetic or katabolic nerve is embryologically connected.[1]

Like other neuralgiæ, angina originates in a lowering of the function of the nerve affected. Usually the nerve function is lowered by long-continued imperfect nutrition, occasionally brought to a climax by some positive cause of ischæmia, as vascular spasm, etc. More rarely the nerve function is depressed by actual pressure upon some of the branches of the cardiac or aortic plexuses, by an aneurism, a tumour, or a dilated aorta. Those conditions which do not necessarily involve ischæmia have always, in my experience, been accompanied by a less severe, though not always a less dangerous, form of angina.

Ischæmia is well known to be a cause of severe pain.

I suppose Jenner was the first to point out the probable connection between ischæmia and angina. He does not explicitly state this connection, but he certainly implies it in saying, " The importance of the coronary arteries, and how much the heart must suffer from their not being able to fulfil their

---

[1] *Vide* Gaskell, *The Journal of Physiology*, Vol. vii., p. 1, and especially, pp. 41 and 46.

function, I need not enlarge upon."[1] As Kreysig has said, "The *word* ischæmia was not then invented, but the *thing* itself was well known."[2] That ischæmia does give rise to pain, even of the most atrocious character, is sufficiently attested by the agony that attends compression of an artery for aneurism, especially at the moment the vessel becomes completely occluded; the pains, arising from a similar cause, that precede the appearance of gangrenous patches in a limb affected with senile gangrene; and those which precede, accompany, and follow attacks of local asphyxia (Raynaud's disease). There is every reason to suppose that the arterial spasm, which is so evidently the cause of local asphyxia, and which takes so prominent a share in the production of an attack of *angina vaso-motoria*, occasionally invades the heart either as part of a general condition, or it may be as a distinctly local affection, and that this is a very possible cause of those anginal attacks where no other seems obvious. For myself I can, however, say that I have never yet seen a case of true cardiac angina in which I have been unable to detect some of the physical signs of dilata-

The most serious forms of angina are those in which the least is to be detected.

---

[1] Letter to Heberden in 1778. Vide Baron's *Life of Jenner*, Vol. i., p. 40.

[2] *Krankheiten des Herzens*, Berlin, 1816, Bd. ii., S. 544.

tion of the heart. It may, indeed, be accepted as a fact, to which I know of no exceptions, that the less there seems to be the matter with the heart the more grave is the prognosis, if the anginous attacks are at all serious.

In angina pectoris, as in other neuralgiæ, we have the presence of a permanent lesion coupled with only occasional attacks. For these attacks there is always some more or less obvious cause. Parry said long ago that the symptoms of angina arise from a temporary increase of weakness in an organ already weakened.[1] Doubtless this is the case when spasm affects the coronaries and diminishes the blood-supply of a heart already suffering from malnutrition. As a rule, however, it is quite the other way; it is not the weakness of the heart, but the work it has to do, that is increased, and the work may be increased in various ways.

Exertion is the commonest cause of increased cardiac action, because the metabolism of the heart and other muscles, when in action, requires more frequent flushing with blood than when they are quiescent, especially if the blood is defective in oxygen or in nutritive material. And exertion after a meal is more apt to induce a paroxysm

Various ways in which angina may be brought about.

[1] Quoted by Stokes, *op. cit.*, p. 486.

than when the stomach is empty; first, because a full stomach impedes and oppresses the heart; and second, because shortly after a meal the vessels are fuller and the blood pressure somewhat raised. *Angina from exertion.*

Any overwhelming emotion may prove suddenly fatal by its action on the heart, and when sudden death has been preceded by repeated attacks of angina, as in the case of John Hunter, it has been assumed that death has been due to this cause. *Angina from emotion.* And this assumption is probably correct though the fatal seizure is often an *angina sine dolore*, an instantaneous death without a cry or any indication of suffering. In such cases the heart may be suddenly arrested in diastole through vagus action, an arrest which the katabolic action of the augmentor nerve fails to overcome; or the emotion may induce irregularity with residual accumulation, which the augmentor nerve fails to expel. In the one case death is instantaneous, as in so many recorded instances (*vide* note, p. 31, *antea*); in the other case, as in that of John Hunter, there may be time to retire to an adjoining apartment before the heart actually fails and death ensues. In both forms death arises from failure of katabolic action, and both may, therefore, be claimed as deaths from angina.

Exposure to cold, especially to a cold wind striking the chest, is a very common cause of angina; the attacks thus brought on are, however, often so slight as almost to be regarded as mere neuralgia — nerve pain — from cold, were it not that relief so immediately follows the use of nitro-glycerine — or some similar remedy — as to make it clear that the chilling of the surface had sufficed to contract the superficial vessels, and so raised the blood pressure as to induce a paroxysm of angina.

Angina from cold.

Occasionally the spasm of the vessels arises not from cold, but from some internal cause; some organic derangement — stomach, liver, or other organ; or from some impurity of the blood; and then we have coldness and numbness of one or more of the extremities, followed immediately by anginous pain.[1] In one such patient the coldness and numbness at first affected the right arm alone, and his heart was fairly good, but this organ dilated considerably before his death, which happened suddenly about two years after he was first seen.

Vaso-motor angina.

Pain, though so usual a concomitant of an attack that angina cannot even be thought of without

---

[1] Landois, *Lehrbuch der Physiologie des Menschen*, 7te auflage, Wien u. Leipzig, 1891, p. 817. Also Eulenberg, *Ziemssens Cyclopedia*, Vol. xiv., p. 48.

bringing up with it the idea of intense agony, yet forms no essential part of the disease, and it is no misnomer to speak of *angina sine dolore*.[1] Most if not all fatal cases are of this character; so far as my experience goes, by far the greater number of fatal seizures have been apparently painless.

In ordinary cases of painless angina there is breathlessness, but no pain, and the attack gets the name of cardiac asthma.[2] A nocturnal attack of cardiac asthma is often the beginning of the end, the earliest indication that the senile heart has become seriously dilated. Now and then ordinary attacks of painful angina cease, and towards the close of life the patient suffers only from fits of breathlessness. At other times the attacks of pain and of breathlessness alternate. And at still other times the pain, with which the attack commenced, passes off and leaves behind it a cardiac asthma as a continuance of the seizure. I myself have assisted at the development of a case of the last mentioned

Cardiac asthma a vaso-motor angina sine dolore.

[1] *Vide* Gairdner in *Reynold's System of Medicine*, Vol. iv., p. 566.

[2] Stokes says: "Well-marked instances of the affection as described by Latham are rarely met with, and the same may be said of the purely nervous cases noticed by Læennec. I have never seen either of these forms. The disease which in this country (Ireland) most often gets the name of angina pectoris might be more properly designated cardiac asthma." — *Op. cit.*, p. 488.

variety. A man aged fifty-seven had suffered from angina occasionally for years; the seizure was brought on by exertion or by emotion, and the pain shot from mid-sternum through to the back and down the left arm, occasionally down the left leg, and sometimes down the right arm also. The heart had a feeble impulse and was slightly irregular, the aortic second was accentuated, and the first sound over the apex was blunt. The radial pulse was tense. Signs which indicated a high blood pressure and a dilatable and probably somewhat dilated heart. While I was listening to his heart sounds, nervous excitement brought on an attack of angina, pain accompanied by a feeling of suffocation. Gradually, as I listened, a distinct auricular murmur developed, and *pari passu* with this the pulmonary second became not only markedly accentuated, but also acquired a booming quality. Evidently residual accumulation, due to irregular and imperfect ventricular contraction, had overdistended the ventricle and promoted regurgitation through the mitral opening, causing considerable pulmonary congestion. From the same cause the radial pulse, which had been firm and tense, became gradually small and feeble. Obviously there had been a transference of the blood pressure from the aortic to the

Case of transference of pressure from the aortic to the pulmonary vascular system.

pulmonary system, and with that an increase of the breathlessness, amounting to a slight attack of cardiac asthma.

No doubt ordinary attacks of cardiac asthma of the kind I now speak of have all a similar origin; irregular or imperfect cardiac action, not marked enough to induce anginous pain, yet avails to induce residual accumulation and mitral regurgitation in a dilated or dilatable heart, and so an attack that begins as a reflex spasm of the systemic arterioles ends in pulmonary congestion and cardiac asthma.

In one old lady the illness commenced with an attack of pain in the *scrobiculus cordis* simulating the passage of a gallstone;[1] her subsequent attacks were simply fits of intense breathlessness without pain, accompanied by a feeble, wobbling heart-beat, and a hard, wiry pulse of high tension. As this tension relaxed and the pulse became soft, the attack passed away. After a very severe attack one evening, which lasted over an hour, she fell asleep, woke in the early morning with a slight attack

---

[1] "It occasionally happens that the very intense and sickening pain of biliary calculus presents a degree of resemblance to angina in its accessories; and the author has even observed cases in which the diagnosis remained doubtful until the yellow tinge of the conjunctiva, appearing after an interval of hours, relieved the apprehensions of the physician." — Gairdner, *loc. cit.*, p. 546.

of Cheyne-Stokes respiration, and passed quietly away.

Many of the victims of fatal angina pass away unobserved during the night; but not a few have had Ishmael's privilege of dying in the presence of their brethren, sometimes suddenly and without warning, while at other times death has been preceded by a longer or shorter period of conscious sinking.[1] So far as my own experience goes, by far the larger number of fatal seizures have been apparently painless. Death has occurred in those fatal cases precisely as it does in animals which have had their coronary arteries artificially blocked;[2] sometimes the heart has failed suddenly; at other times complete

Death from angina generally painless. The fatal asystole may be sudden or ingravescent.

---

[1] In one recorded case this conscious sinking occupied quite half an hour. Vide Balfour, *op. cit.*, p. 305.

[2] *Vide* Panum, *Virchow's Archiv*, Bd. xxv., 1862, S. 308, etc.; Von Bezold, *Untersuchungen aus dem physiologischen Laboratorium in Würzburg*, erster Theil, 1867, S. 256, etc.; and See, *Comptes Rendus*, Tome xcii., 1881, p. 88. See found that section of the vagus did not in any way modify the phenomena produced by occlusion of the coronaries, and also that stimulation of the vagus had no effect whatever on an ischæmic heart. Complete ischæmia is not always immediately fatal, because of the enormous reserve of energy that the heart possesses, which it takes some time to exhaust. But a long continuance of imperfect nutrition modifies this reserve of energy in a most important manner, diminishing it remarkably. — Von Bezold, *loc. cit.*, S. 279.

failure has been preceded by a longer or shorter period of ingravescent asystole (more or less conscious sinking), — the longer or shorter time occupied in dying evidently depending upon, first, the degree of ischæmia, actual or relative, present in each case, and, second, the length of time during which comparative ischæmia has already persisted, and the consequent amount of exhaustion which the cardiac energy has already sustained. The pre-existing nutritional vigour of the heart, and the nature of the exciting cause, have all somewhat to say in regard to the actual mode of death. One gentleman, over eighty years of age, who had long suffered from angina, took his seat at a public meeting, and, without a sigh, sank down dead. I have just mentioned (p. 138) one case in which ingravescent asystole occupied fully half an hour. When I entered this patient's bedroom, he said to me, "Doctor, this is very different from anything I have had before," and he died quietly, after drinking about half a glass of brandy given him in the hope of stimulating his heart to more vigorous contraction. A few years ago I saw an old gentleman for severe angina. Some weeks subsequently he assisted home a friend who had met with a slight accident; this made him feel very unwell, but he struggled to reach his own home, took to bed, and died within twelve hours, never having recovered from his exhaustion. Another old gentleman, long a sufferer from angina, but who for the last two years of his life had been free from pain, a week before he died had what was called a faint, really an attack of cardiac failure, — an *angina sine dolore*. During the week for which his life was spun out by the judicious use of stimulants and a careful avoidance of the most trifling exertion, he had several trifling attacks of a similar character. The final seizure was absolutely painless; life ceased because the heart failed to contract.

Again, during last winter (1892–93) there was a man in Chalmers Hospital suffering from aortic regurgitation, accompanied with attacks of angina, with a tense pulse and high blood pressure, evidently of reflex origin and chiefly occurring at night. After much suffering, this man's strength at last broke down; but he took many weeks to die, and his death was a most notable example of ingravescent asystole (*vide antea*, p. 80). The degree in which the cardiac energy has been previously exhausted by malnutrition — imperfect metabolism — regulates the rate at which asystole progresses. Long-continued malnutrition of a serious kind, rapid failure at the last; less serious interference with the metabolism of the heart, involves a slower process of dying. At times we can predict the near approach of death, but in the greater number this is impossible, and chiefly for this reason, that, even when the cardiac output is greatly diminished, the blood pressure does not fall *pari passu*, but for a time remains normal till the fatal limit is reached, when the pressure suddenly falls, and death ensues.[1]

Death from angina is thus not always instantaneous, nor is angina always fatal. Angina may even be recovered from, sometimes perfectly, for when the cause is remediable, the angina is also curable. At other times the pain is removed, but the disease progresses, and after a longer or shorter period free from suffering, or with only occasional attacks of cardiac asthma, the patient at last succumbs, dying of angina, no doubt, *sed sine dolore*.

Elements of prognosis in angina pectoris.

In endeavouring to get a basis for a prognosis in any case presenting symptoms indicative of angina, the first point to ascertain is whether we are dealing with a true angina, or merely with one or other of the varieties of neuralgia referred to at page 117.

In the absence of any opportunity of observing a seizure, there may occasionally be some difficulty

[1] *Vide* Roy and Adami, *British Medical Journal*, December 15, 1888, p. 4 of the reprint.

in deciding this matter. Any indication, however slight, of dilatation of the heart must be looked upon as a point in favour of the reality of the angina, and this at any age. In youth the probability is greatly in favour of this dilatation depending on curable spanæmia, and therefore of the angina itself being curable. After middle life we may still have to deal with a curable spanæmia, though the dilatation in no case depends solely on this; and the next point to determine is, have we here simple senile dilatation, or are the coronaries also in any way obstructed? Atheroma of the external arteries affords a certain presumption in favour of the coronaries being also atheromatous; but even if they are, it is obstruction and not atheroma that induces ischæmia, and the question at issue can only be decided by the results of treatment.

On the other hand, if there be no indication of cardiac dilatation, — which is but seldom the case, — an anxious, haggard expression unmistakably indicates great suffering and a serious disease, while a countenance free from distress or anxiety is an equally certain indication that, whatever may be wrong, we have not to deal with a serious angina. But though not due to angina, substernal pain may be quite as dangerous : it may be due to a small aneurism impossible to detect. Such cases are great sources of anxiety to the physician, and too great caution cannot be exercised in giving any decided opinion in regard to them.

The large reserve of energy with which every heart starts in life enables it long to resist the many injurious influences to which it may be exposed; hence, considerable enlargement is quite consistent with perfect freedom from angina, if the arteries are pervious and the blood of good quality; yet such a heart has its metabolism easily upset, and comparatively slight causes suffice to induce an anginous seizure. There is little wonder, then, that so many senile hearts are subject to angina; the wonder is rather the other way, — that there are, comparatively, so few hearts, even of those showing decided signs and symptoms of cardiac degeneration, that are affected by this complaint. Statistics are proverbially uncertain, and the character of any man's experience depends very much on where he has obtained it, so I give mine only for what it may be worth, without claiming for it any special accuracy in regard to any one particular.

Confining myself solely to those patients who consulted me at my own house, during the ten years 1879–89, I find that I have notes of 1173 cases of various affections of the heart and aorta. Of these, 581 were senile hearts, 270 were young spanæmic hearts, and only 57 were cases of heart affection distinctly traceable to rheumatism. Of the 581 senile hearts, 98, or rather over one-sixth, made a prominent complaint of angina. Of these, 17, or nearly one-sixth, were females,— a considerably larger proportion than the 8 females out of 88 cases collected by Sir John Forbes, but the number is still sufficiently small to make it probable that Forbes is right in saying that angina is more common in men than in women; a proposition which few will be inclined to deny.[1]

Relative proportion of senile to other diseased hearts, and of angina to the senile hearts.

Of the 98 cases of angina, 15 are certainly known to have died, without having gained anything from treatment beyond palliation of the symptoms; and 17 are known to have got entirely free of their painful seizures. Of these 17, 13 are still alive in apparently excellent health, and 4 have died after a longer or shorter period of complete freedom from pain.

One of these patients was for four years and a half completely free from pain; during this period

---

[1] *Cyclopedia of Practical Medicine*, Vol. i., p. 83. But Forbes adds: "Of milder cases, a very considerable proportion, perhaps an equal proportion, are met with in females, and at an earlier period of life. This at least is the result of our own experience; and the same opinion is entertained by writers of great authority." — *Loc. cit.*

he progressed from sixty-nine to seventy-three years of age, and was able to carry on his business, attending markets in all parts of the country. At last, after attending market in the country town in which he lived, and having transacted business in apparently perfect health, he returned home, sat down, and without a complaint quietly departed. When first seen, this gentleman had suffered from angina for eight weeks, and attacks came on whenever he attempted to go up any ascent, and the pain extended into both arms, but chiefly into the right one. For years he had been breathless; his arteries were all atheromatous, hard, and tortuous; his heart was dilated, with a feeble impulse, and there was a loud systolic murmur in all the cardiac areas. Under treatment, the pain entirely ceased in a few months, the heart's impulse became much stronger, the murmurs less distinct, and he declared he was as able as ever to go up hills or stairs. About seven months before his death he had a severe fall, and was never so well afterwards. On the day of his death he had been about his business all day, apparently as well as ever he was; he went out again in the evening, returned about half-past seven, sat down, and quietly died. After death his face was pale, with a most placid expression, his pupils were both nat-

Illustrative cases.

ural. There was no post-mortem examination of the body.

A second case was that of an old gentleman, between seventy and eighty years of age, who was two years under treatment before he got rid of his pain. The remedies employed gave great relief, but the pain continued to recur upon exertion, and sometimes in bed if his stomach was flatulent, for quite two years; after this he had no more attacks of pain. When first seen this patient had a feeble impulse, and a loud systolic murmur in all the cardiac areas; when last seen, just a week before his death, his apex-beat was firm, the systolic murmur loud, and the heart's action intermittent and irregular. His arteries were very hard from the first. About four years after he was first seen — two years after the cessation of pain — he had what was described as a bad faint; two days subsequently he had another slighter attack of a similar character, but no pain; and just one week afterwards he died quietly and suddenly one forenoon, sitting in his chair.

The third case was that of a fresh-looking man of sixty-one, who had firm arteries with high blood pressure, a large, dilated, and somewhat hypertrophied heart, with considerable palpitation and irregularity of action. He had severe anginous pain across his chest upon exertion, extending down the right arm. The urine was of low specific gravity, — renal inadequacy, — and there was an occasional trace of albumin. He had also long suffered from an irritable bladder and an enlarged prostate gland. The angina was speedily relieved by treatment; but about four months subsequently he had an attack of pneumonia of no great severity, and during convalescence he died somewhat suddenly from uræmic sinking.

The fourth case was that of an old gentleman between seventy and eighty, who, after nearly ten years' relief from pain, and of really excellent health, was suddenly seized during the night with an attack of cardiac asthma and died in about half an hour.

Of the other thirteen who got rid of their attacks of pain, some have been free for quite ten years, others for varying periods down to five years. Most of them I see occasionally; the rest I hear of, and I know them to be well, each with a good firm heart-beat and no pain on exertion; even breathlessness is not much complained of, though where murmurs did exist they still persist. The great difference between their past and present is that whereas their hearts were formerly feeble and ill fed, they are now strong and well fed. To keep them in this condition they require constant care, watching, and treatment, and in the face of advancing age none of these can be long pretermitted without risk of a serious relapse.

Of the fatal cases to whom treatment gave only temporary relief, two were little over middle life; both were busy men. In the first of these the apex beat below the left nipple; the first sound was almost absent, quite faint and impure; the aortic second was accentuated; there was no distinct murmur anywhere detectable; the fits of angina were very severe and easily brought on. I saw this patient in one of his attacks, and formed a most serious prognosis from the severity of the pain, the comparative youth of the patient, and the little that was to be found wrong with his heart. He dropped down dead in his own hall about a month after I saw him.

The second patient had a slight systolic murmur in the mitral area; the apex beat just inside the left nipple; the aortic second was accentuated. The fits of angina were said to be very severe, but I had no opportunity of seeing one. I gave an unfavourable prognosis for reasons similar to those given in the former case. Three months subsequently this patient was found dead in his office.

Three patients had serious illness connected with the heart for some weeks before death. One of these was never well after his first attack of angina; he had a large, dilated heart, and died within two years. The other two were country lawyers, who carried on their arduous business, one of them for ten and the other for seven years, subsequent to their first attack of angina;[1] both died after short illnesses following overexertion in the course of business.

One clergyman, after suffering for about seven years from a dilated heart with angina on exertion, hurried to catch a train at a country station, and died resting in a chair on which he sat down exhausted when he reached his goal.

Another clergyman had been out for a drive one bleak November day; on coming home he sat down by the fire, complaining of cold, and slipped from his seat dead. He was sixty-five years of age, and had a large, dilated heart, with a shrill systolic murmur in all the cardiac areas. He died within a month of being seen, having been much relieved by treatment.

A third clergyman (each of these was of a different persuasion) wrote me as follows a few weeks before his sudden death: "I don't know whether

---

[1] As I was writing this, the son of one of these lawyers, a man thirty-eight years of age, called, complaining of anginous pains, on exertion, of short duration and not severe. His heart was weak, its sounds parchmenty in character, the blood spanæmic from imperfect recovery from an attack of drain-poisoning a year ago. He recognized the pain as similar to what his father had suffered from. He died within two years.

you will remember my consulting you about a year ago. . . . As long as I keep still I have no discomfort, but very frequently, though not always, if I walk fifty or sixty yards I am seized with the most painful spasms, not in my heart, but in the pit of the stomach. If I stand up, the pain comes on; even putting off or on my clothes excites it. Yesterday I went to my garden, a distance of less than one hundred yards; I had severe pain, and on coming home our local medical man, who happened to be in the house, found my pulse to be intermittent; this must have come on recently, as he never observed it before. I am seventy-one years of age, and with the exception of this pain I am in perfect health, and feel as strong and well as I was twenty years ago. As I cannot walk, I ride or drive, and I can do both in moderation, provided I enter my carriage or mount my horse slowly, but any sudden or quick movement brings on discomfort. Sometimes I am for a week quite well, and then without any cause which I can trace I am suddenly plunged into the most extreme discomfort. A brother of mine suffered for some time from exactly the same symptoms; he was told he was suffering from long-standing heart complaint. For the last three years all his symptoms have disappeared, and he is now quite well." This old clergyman had hard, atheromatous arteries, a large, dilated heart, with a systolic murmur in all the areas, and the aortic second so feeble as to be quite inaudible. No diastolic murmur was to be heard. He died quite suddenly a few weeks after writing the foregoing letter; no treatment gave him any relief.

Other four angina patients of the ninety-eight referred to also died suddenly. One of these was a case of free aortic regurgitation, with a large heart; the other three were cases of dilated heart, two of them with a blunt first, an accentuated aortic second, and no murmur; the fourth had a large, dilated heart, an accentuated aortic second, and a systolic murmur in all the areas. He was fifty-eight years of age, and treatment gave him great relief, but he died about four months after being first seen. One morning after breakfast he attempted to raise himself in bed, gave a few gasps, and died. I have no particulars as to the mode of death in the three preceding cases.

The following case is very instructive from more than one point of view. On the 21st of January, 1880, I was asked to see a gentleman suffering from angina. I found him to be a well-preserved man of sixty-eight, suffering so much that the exertion of walking even ten yards, or taking off or putting on his clothes, sufficed to bring on an attack of pain, distressing enough, but not of great severity. His heart was slightly enlarged; the apex beat in the fifth interspace almost directly below the left nipple; the first sound was blunt, the second accentuated, the arteries hard and atheromatous. I prescribed for him, and recommended him to get out of business, and if possible into a warmer climate. As he had already suffered much from physicians, I gave him, at his own request, the following letter, to show to any physician who might be called in, the view taken of his case, and the lines on which it was desired to have the treatment carried out.

JANUARY 21ST, 1880.

DEAR SIR: I shall group what I have to say under three heads:—

    1. The nature of your disease and its cause;
    2. The treatment, medical and general; and
    3. The results to be expected from that treatment.

1. The cause of your disease is primarily a loss of elasticity of the arteries; by this a greater strain is thrown upon the heart than usual. The result of this extra strain is, in your case, a slight enlargement of the heart — dilatation with compensatory hypertrophy. In your case the hypertrophy is somewhat insufficient, and the result is that when your heart is called upon for any unusual exertion, either by emotion or bodily exercise, it becomes pained, and the pain shoots along the course of various nerves connected with those of the heart, and thus gets referred to various other parts, as the arms or stomach, after a fashion with which

medical men are only too well acquainted. In its essence it is an *angina pectoris*, not associated with a lesion of any valve, though I quite believe that the signs of mitral regurgitation may sometimes be present.

2. The medical part of the treatment resolves itself into means to relieve the pain when it occurs; and, secondly, the use of remedies to improve the condition of the heart, and thus lessen the frequency of the attacks. These prescriptions will be found on a separate paper, and on it the medical man, whom you must always send for at once on the occurrence of any seizure, will also find noted down suggestions as to certain supplementary measures which may be employed if the attack does not quickly yield to the means first employed. The general treatment must consist in a persistent avoidance of all bodily or mental excitement. You should, therefore, go about on foot as little as possible; at present drive only. Shun all public meetings and worry of every kind. To gain this complete rest of body and mind, as well as to escape our often severe cold, which for you is not devoid of danger, I would strongly recommend you to start at once for the south of England at least, and if at all able, for the south of France. I have no doubt that you will benefit greatly by the change, and I think you will manage it comfortably, taking the journey by easy stages.

3. The amount of benefit you are to expect from treatment depends upon conditions as yet unknown to me, but which will reveal themselves by and by. What I aim at is to put on the drag, so as to stop you from going down hill so rapidly as you have been doing. If your heart muscle is as sound as I believe it to be, the treatment may result in completely stopping the pain; more probably, from the long time the pain has already troubled you, it will only come seldomer and be less severe. The causes of your ailment are incurable; we can only mitigate the results. Sometimes the relief obtained is so great as to simulate a perfect cure, and this is what we aim at. I am, etc.

The following December I wrote the patient's doctor to enquire as to his condition, and received the following reply: "I saw Mr. B. about two months ago, and found him in excellent health. I believe he is at business every day, and able for a good day's work. After you saw him with me in spring he went to Bournemouth, with which he was delighted. He stayed there three months, and came home in June immensely improved in every way. No return of the angina pectoris, nor any tendency thereto, since you last saw him. July and August he spent at Crieff, and thought this also helped him greatly. When I examined him in June, there was still evidence of the dilatation of the heart, but I could make out no murmur. His breathing power was also much improved; he could walk much farther, without ever requiring to stop as formerly."

Early in the following March Mr. B. paid me another visit; he assured me that so far as he could judge he was as well as ever he was, and that he had only called to obtain my sanction to his marriage, as he thought an agreeable companion would greatly conduce to his comfort and happiness.

I found the heart's impulse much improved in strength, otherwise the condition of the heart was as formerly, and my sanction to his marriage was given along with some sage advice. After this

Mr. B. continued in the enjoyment of apparently perfect health for about two years. Then his partner in business died rather suddenly, and as this partner had for some years taken the larger share in conducting their very extensive business, the whole of this, with all the correspondence, was suddenly thrown upon Mr. B. The result was a complete breakdown; the work and worry were too much for his heart; this organ rapidly dilated, it developed a loud systolic murmur in all the areas, its action became embarrassed, very considerable general dropsy set in, and the case looked most serious. Fortunately, the recuperative power of the heart was not lost; it responded well to treatment, and a few weeks of perfect rest restored Mr. B. again apparently to his former state. His health seemed quite re-established, he suffered no more from his heart, and he was able to go about and enjoy himself, though debarred from any longer taking an active share in business. Two years more passed away, and again I saw Mr. B., this time for a slight paralysis affecting the left arm, apparently due to some cortical thrombosis or small embolism. This paralysis came on with giddiness as he was leaving church one Sunday afternoon; it speedily passed away, and left him not a whit the worse. Six months afterwards he died from pneumonia. Being from home, I did not see him upon this occasion, but I understand that his damaged heart undoubtedly hastened his end. It is the rarest thing in the world — if indeed it ever happens — for a man over sixty, with a dilated heart, to recover from pneumonia. Even if he recovers from the primary attack, the exhaustion following it initiates the beginning of the end, which is no long time in following.

CHAPTER VII

THE SENILE HEART, ITS CONCOMITANTS, AND SEQUELÆ. GOUT

LATHAM has finely said that the clinical history of diseases of the heart is but the history of "those prior and accompanying conditions in the life and health of the patient, which were found variously leading to and variously promoting and causing them; as well as all those subsequent conditions in the life and health of the patient variously springing *from* them and variously promoted and caused *by* them." And Latham adds that the treatment of such diseases is but the employment of "means of influencing those same conditions and of influencing them for good."[1]

These statements — so true of all cardiac diseases — find their pre-eminent application in senile affections of the heart. For the clinical history of these affections is not the mere story of the past few weeks, but comprises the life-history of

[1] Latham, *op. cit.*, Vol. ii., p. 360.

the patient from his cradle to his grave, and often includes that of his forefathers also. They form part and parcel of the patient's development, and are the natural result of prematurity, or of excess, in those changes affecting the arterial tissue which wait upon advancing years, and which in their turn variously modify every subsequent condition in life. These tissue changes commence long previous to the time when either symptoms or organic changes force themselves upon the attention of patient or physician. During all this time they have been slowly but persistently exerting a modifying influence, not alone upon any one organ of the body, but on every tissue of which the body is composed, as well as on every function which its organs discharge.

The story of senile cardiac disease comprehends the whole life-story of the patient from his cradle to his grave.

The arterial dilatation, already referred to, which results from the loss of elasticity, coupled with lessening of the capillary area from obsolescence of many of these vessels, gradually induces another change in the circulatory system too often forgotten or overlooked, but which is yet a factor in every function of our future life important enough to demand our most serious attention.

Up to the completion of puberty the pulmonary artery is larger than the ascending aorta; with the advance of maturer years, and the coincident changes in the circulatory system, a change takes place in the relative size of these vessels.[1] The effects of this change are of the greatest importance. So long as the pulmonary artery remained the larger of the two, the blood within the pulmonary circulation was kept at a high pressure, and this notwithstanding a free and unembarrassed egress into the left heart, and so onwards. But blood circulating through the lungs at a high pressure, and at the normal rate, gets rid of its carbonic acid more rapidly and also more perfectly than blood circulating at the same rate but at a lower pressure. Hence up to nearly middle life every tissue of the body has been continuously flushed with a highly oxygenated blood, full of potentiality, and a well-nourished and healthy organism has thus been filled full of life and vigour, and placed at its very best in regard to capacity for bodily and mental exertion. After the full development of puberty all this slowly changes; under the influences already de-

Change in the relative size of the pulmonary artery and the aorta takes place about middle life.

Effects of a high intra-pulmonary blood pressure in health.

---

[1] *Vide* Beneke, *Die Altersdisposition*, Marburg, 1879, ˙S. 18. Also "Ueber das Volumen des Herzens und die Weite der Arteria pulmonalis und Aorta ascendens in den verschiedenen Lebensaltern." In *Schriften der Gesellschaft zur Beförderung der gesammten Naturwissenschaften zu Marburg*, Cassel, 1879, S. 5.

tailed (*vide antea*, p. 11, etc.) the calibre of the aorta becomes gradually greater than that of the pulmonary artery. The result of this relative diminution in the size of the pulmonary artery is that the blood circulates through the lungs at a much lower pressure than formerly, the carbonic acid is consequently given off more slowly, and throughout all the future life there is a gradually increasing "venosity" of the blood, as the older writers called it, which has an important influence on every function of the body. As age advances there is also a slowly increasing tendency of the blood to accumulate in the veins at the expense of that contained in the arteries, and the slightest disposition to cardiac debility aggravates this tendency. The result of this is an increasing disposition to venous congestion, to remora of the blood in the venous radicles, and also to accumulation of the serous plasma in the extra-vascular spaces. The influence of the increased tension thus produced within these spaces upon the intra-vascular blood tension, and through that upon the heart, has been already referred to (*vide antea*, p. 28). We shall presently see that this extra-vascular remora of the blood-plasma also forms an element in one very important disease, and often gives rise to local morbid phenomena of an interesting if not dangerous character.

Thus, after middle life, the blood is being continually shut off from ever-increasing areas throughout the body by withering of the capillaries; it slowly accumulates in the veins, it is less highly oxygenated than formerly, and is thus less fitted for promoting the discharge of any of the vital functions. *Effects of these vascular changes on the blood.*

But this condition of "venosity" of the blood, and of remora of the blood in the veins, — "venous congestion," as it is so usually termed, — has been recognized by all physicians since the days of Galen as "the first condition essential to the formation of the gouty diathesis."[1] It is the basis of that diathesis.

Add to this that in a state of civilization man is always supplied with a superfluity of foods and drinks, which the habits of society and the anxiety of his friends tempt him, if they do not actually compel him, to partake of four or even five times a day. *The alterations in the blood resulting from these vascular changes are the basis of the gouty diathesis.*

Moreover, as the bubbling energy of youth fails, the mere pleasure of it no longer incites us to violent exertions, the needs of civiliza-

[1] *On Gout.* By W. Gairdner, M.D., London: 1849, p. 121.

tion do not require such exertions from us, and the many luxurious appliances of civilized life aid and abet the natural indolence that grows upon man as age advances, and largely preclude the need for any but the most trifling bodily exertion.

Hence this less highly oxygenated blood is flooded with a redundancy of nutritive material, far in excess of the requirements of the frame, which can neither be used up in any of its ordinary appropriations, nor fully oxidated in any other way, and so excreted. The general metabolism is thus impaired, every function of the body impeded, every secretion deteriorated; all the organs suffer.

Thus we have the Gouty Diathesis fully developed; a diathesis — habit of body — present in each one of us after middle life, and which modifies the organic metabolism of each one of us, both in health and in disease. The gouty diathesis is only a comprehensive term for all those changes in the character and composition of the blood induced by the evils of civilization — deficient exercise and excess of nutriment — multiplied into those developmental changes in the vascular system, which are at once the cause and also the consequent of puberty.

Gout, on the other hand, is the name given to

all those modifications of our metabolism caused by the gouty diathesis, as well as to all the symptoms to which those modifications give rise. Naturally, after middle life gout affects every organ of the body, both in its structure and its function; in a state of civilization we are all, after puberty, more or less gouty, and we are gouty in a gradually increasing ratio.

Gout only a generic term applied to those modifications of metabolism caused by the gouty diathesis.

A paroxysm — painful and distressing though it be — is a mere episode in the history of gout, and an episode indeed to which not a tithe of the gouty are liable. The severity of the symptoms are but an accident of locality; and the pain like that of angina pectoris itself is but the product of ischæmia.

A paroxysm of gout is a mere episode in its history.

It is many a year ago since Cullen rejected the term *Arthritis* as inapplicable to the gouty paroxysm, because it hinted at an inflammation which had no existence, and gave it the name of *Podagra*, as expressive of the one fact most undeniable about it, that though it may occasionally affect other parts, it is most usually a severe pain in the foot.[1] Cullen's description of an attack is still as accurate as ever, and the appearance of the

[1] *Synopsis Nosologiæ Methodicæ*, Edin., 1815, p. 17, note.

part affected is the same now as then, yet we seem to have made but little progress in discovering its true nature.

If there is one thing more certain than another about an acute gouty affection of a joint, it is that, though often regarded as an inflammation, it presents none of the characteristics of that process, and ends in none of its usual terminations. There is never any suppuration, and never any adhesion of the opposed surfaces. There is here no process of abnormal nutrition, no cell-proliferation, no diapedesis of the white cells, no true inflammation.

Distinctive marks of a gouty joint.

Never any adhesion or suppuration.

It may be objected that a gouty joint always ends in resolution, and that resolution is one of the natural terminations of inflammation. But the resolution of a gouty joint is always incomplete; even after a first attack there is left behind a deposit of urate of soda in and around the joint, and this increases with each subsequent attack, so that a gouty joint never returns to its pristine condition, but gets larger and stiffer with each attack.

Resolution always incomplete.

Again, of the well-known *Quatuor Notæ*, the *Calor* is often, if not always, wanting; the temperature of the affected joint may even be below the normal. "I

Calor wanting.

have found it 97, while that in the mouth at the same time was 100."[1] This is quite incomprehensible except on the supposition that there is some obstruction to the free circulation of the pyrexial blood through the part affected, and indeed the whole history of a paroxysm is most readily explicable upon this supposition, while at least one kind of treatment is quite inexplicable upon any other.

A paroxysm begins with a sudden attack of acute pain, which may pass off as suddenly as it came, leaving the joint unaltered; or the pain may increase, and become excruciating; the joint swells, becomes dusky red, tense and shining, and the veins leading from the joint to the dorsum of the foot are dark and turgid. The attack generally begins with slight shivering; the pain is compared to that of a penal boot or thumb-screw; this torture, made unbearable by the slightest vibration, lasts till morning — till cock-crow, *galli cantu*, as Sydenham puts it; then slight remission takes place, the patient falls into a gentle perspiration, and at last gets to sleep. In the morning the joint is found swollen, shining, dusky red, and the pain is easier. This remission lasts through

History of a paroxysm.

Dark turgid veins lead off from the part affected.

---

[1] *Vide* Sir Dyce Duckworth's *Treatise on Gout*, London, 1889, p. 248.

the day, but towards evening the pain recurs as severe as ever, and this cycle of remission and exacerbation goes on for four to eight nycthemera. Then the crisis is over, the remissions get gradually longer and more complete, and erelong there is nothing left but a numbness which may last for about a week. The redness of the joint attains its maximum intensity in about thirty hours; it then diminishes or rather gets more violet in hue as the pain wears away. The œdema increases for four or five days, and when it disappears and the attack is over the joint remains stiff, the foot soft and numb, and the gait hesitating for other ten or fifteen days.

Paroxysm may last from a fortnight to three weeks.

Throughout the whole history of a paroxysm there is no indication whatever of inflammation; there is no known inflammation which runs so fixed and definite a course, and so invariably terminates in resolution.

A paroxysm of gout not an inflammation.

On the other hand, if we accept the idea of the gouty joint being an *infarction*, the phenomena are easily explicable, and the invariable termination readily accounted for.

Infarction suggested as an explanation of the condition of the gouty joint.

An infarction is the gorging of a part with serum, blood, or both; it presupposes a block in the circulation, the formation of an anæmic area, and the gorging of this area with retrograde blood from the neighbouring valveless veins.[1]

There is no difficulty in imagining the occurrence of a block in the circulation of any gouty person, as we know gouty thrombosis to be in them a very common occurrence. Recumbency for a few days for some trifling ailment is quite sufficient to induce thrombosis in one or more of the veins of the extremities in many, necessitating three weeks longer in bed than was bargained for. In others, some unusual sedentariness of occupation is quite sufficient to cause thrombosis, which gives rise to no pain unless it be connected with some tendinous part such as the heel, where indeed it is but slight and evanescent, as it sometimes is when it occupies but a limited area even in the usual point of selection, the junction of the metatarsal bone with the proximal phalanx of the great toe.

Granted the block in the circulation, then the other phenomena follow in regular sequence as a matter of course. Arrest in the onward movement of the blood in the veins leading from the part is speedily followed by their turgescence, because the blood flows into them from the sur-

[1] Cohnheim, *op. cit.*, p. 121.

rounding capillaries and valveless veins until an equilibrium is established between the pressure in the occluded area and that within these veins.[1]

The sluggish movement of the blood also permits the accumulation of the red corpuscles within the capillaries of the affected area, hence the red turgescence of the part, a redness that grows duskier the longer it continues.

Moreover, the remora of the lymph within the tissue interspaces not only contributes to the tension of the part, but the lymph being highly charged with the somewhat insoluble salts of uric acid, these may crystallize out during this delay, or they may get left behind as a residuum when the lymph gets reabsorbed as recovery progresses, and thus originate those deposits of urates in and around the joint which increase with each subsequent attack.

The pain is in the later stages largely augmented by the increased tension within the part, but primarily it is due to ischæmia. An acute twinge is often the earliest indication of an attack, this goes on increasing if the other signs are superadded, but the pain passes off at once should the circulation rapidly return to its normal, as it not infrequently does.

[1] Cohnheim, *loc. cit.*

In favour of the idea that a gouty paroxysm is due to a local infarction, we have thus, —

First. A gouty joint contains no inflammatory exudation, but merely a sero-sanguinolent effusion in and around the joint; and neither in its commencement, course, nor termination does it correspond with any known form of inflammation.

Second. In the gouty diathesis thromboses are common enough; they occur in circumstances and under conditions precisely similar to those in which we have a paroxysm of gout evolved, but a gouty paroxysm never follows unless the thrombosis happens to occupy a position in which it necessitates the formation of an infarction.

Third. When a thrombosis occurs anywhere, the time occupied in recovery is precisely the same as that usually required for recovery from a fit of gout. I well remember an old friend who was subject to repeated attacks of gouty aphasia due to cerebral thrombosis; in him the period that elapsed before the power of speech returned was three weeks, — precisely that usually occupied by an ordinary fit of gout.

Fourth. The acceptance of thrombosis followed by infarction as an efficient cause of the gouty paroxysm, not only affords a reasonable and sufficient explanation of all the concomitant phenomena of a fit of gout, but it also supplies a rational and intelligible explanation of a mode of treatment which has proved highly successful, and which is utterly inexplicable in any other way.

When Boerhaave in Section δ of his 1275th aphorism talks of the cure of gout being carried out, "*Exercitio magno, continuato equitationibus in œre puro, tum frictionibus, motibusque partium sœpe iteratis,*"[1] it is quite probable from the context that he refers to the cure of an acute attack, and not merely to massage and other forms of friction as employed to remove the rigidity of gouty limbs.

Acute gout may be successfully treated by massage alone.

At all events we know that Sir William Temple — who was ambassador at the Hague when Boerhaave was born — was aware of this method of cure, for he says that in one part of the East Indies, "the general remedy of all, that were subject to the gout, was rubbing with hands; and whoever had slaves enough to do that constantly every day, and relieve one another by turns, till the motion raised a violent heat about the joints where it was chiefly used, was never troubled much, or laid up by that disease."[2] Temple also

---

[1] *Aphorismi de cognoscendis et curandis morbis*, ab Hermanno Boerhaave, ed. 3tia, Lugdun, Batavorum, 1727, p. 312.

[2] Boerhaave was born in 1669; Temple retired from the embassy in 1671. *Vide* An Essay on the cure of gout by Moxa, in *The Works of Sir William Temple, Bart.* Edinburgh, 1754. Vol. ii., p. 127.

says that the Rhyngrave, whom he knew very well, never used any other remedy for the gout, to which he had long been subject, except on the first indication " to go out immediately and walk, whatever the weather was, and as long as he was able to stand, and pressing still most upon the foot that threatened him; when he came home he went to a warm bed, and was rubbed very well, and chiefly upon the place where the pain began. If it continued or returned next day, he repeated the same course, and was never laid up with it; before his death he recommended this course to his son, if ever he should fall into that accident."[1] Temple also tells of one of his brother's gamekeepers who when seized by a fit of gout never laid himself up, but walked after his deer or his stud from morning to night, in spite of the pain, till he got ease.[2] This reminds us of the statement by Mr. Apperley — the well-known Nimrod of old days — that a friend of his when threatened with a fit of gout, after a good dinner and his *quantum suff.* of wine, warded it off by walking the soles of his pumps quite through before going home to bed. Dr. Gairdner also relates the case of a friend of his own, an old gentleman of eighty-five, whose constant remark to his physician and his family when he was seized with a fit of gout, was, " I'll walk

[1] *Loc. cit.*   [2] *Loc. cit.*

it off"; and walk it off he did. This same old gentleman often quaintly remarked to his friends, "Go to bed with the gout, and it will surely go to bed with you, and be mighty bad company"[1] — a statement which curiously resembles that by Temple that sufferers from gout carry it presently to bed, and keep it safe and warm, and indeed lay up the gout for two or three months, while they give out, "that the gout lays up them."[2]

In the beginning of this century a namesake of my own, apparently quite unaware that anything had ever been written in regard to the treatment of acute gout by friction, wrote a paper on what he called a "New, simple, and expeditious method of curing gout,"[3] advocating massage for this purpose. He narrates three cases in which this treatment was perfectly successful. One of these patients at first rejected the treatment as entirely inapplicable to him, as he had attempted to touch his own toe, and he might as well have applied "living fire." Nevertheless, firm pressure and

---

[1] Gairdner, *On Gout*, London, 1849, p. 114.
[2] *Op. cit.*, p. 128.
[3] By W. Balfour, M.D. *Vide Edinburgh Medical and Surgical Journal*, Vol. xii., p. 432. And none of us have forgotten the old gentleman in *Sandford and Merton* who was cured of his gout by being starved and locked up in a room without a seat, the floor of iron being gradually heated till continual movement became a necessity.

friction entirely removed the pain in ten minutes, and in two days he was going about as usual.

Facts such as these are worthy of the most careful consideration, and are only explicable on the theory that the essential lesion in an attack of acute gout is the formation of a thrombosis — at first probably a mere stasis — in such a situation that an infarction is a necessary consequence.

In the gouty — that is, in all of us after middle life, more or less — thrombosis is always a possible occurrence, and it plays a manifold rôle, the importance in each case depending on the position of the block.

Thus thrombosis of the cortical vessels plays a notable part in progressive softening of the brain; the symptoms varying according to the part affected. Thrombosis or stasis in the motor areas is often limited in extent and temporary in character, so that the resulting paralysis may be slight and evanescent, or more extensive, more complete, and permanent. The same may be said of Aphasia, — so common in the gouty, — which may be either amnesic or motor in character; and either incomplete and temporary; complete and yet temporary; or both complete and permanent; the last rare in purely gouty cases. Incomplete attacks of a paralytic char-

Gouty thrombosis plays a manifold rôle.

Various symptoms from cerebral thrombosis.

acter may be but slight and pass off rapidly, like the twinges about the heel and toes which quickly vanish. Or these attacks may be more complete and yet quite temporary in their nature, lasting just about the usual time of a gouty paroxysm — from two to three weeks. Single attacks of cortical thrombosis are seldom of much consequence, but by frequent recurrence they may ultimately produce most serious results. On the other hand, central thromboses are always most serious; sometimes one or both pupils may be dilated; the breathing may be deep and regular, like the blowing of a bellows; and death may occur in a few hours, preceded by a considerable rise of temperature, coma, or muttering delirium, and sometimes by convulsions.

Venous thromboses of the limbs are more troublesome than dangerous, in those who are otherwise healthy. But as there are many to whom a day or two of recumbency always means the blocking of one or more of the veins of the extremities, even a trifling catarrh to them means three weeks more of bed than it does to ordinary people. It is only when such a patient is out of health that micro-cocci invade these clots, which then break down and give rise to showers of poisonous emboli, producing blood-poisoning of a serious character with scattered

Thromboses of limbs.

septic abscesses. Nay, it sometimes happens, especially in connection with a dilated senile heart, that perfectly aseptic thrombi in some of the smaller veins soften and break down into showers of minute emboli, with no other result but a sudden rise of temperature, putting on the appearance of an ague of irregular type, and prolonging convalescence till blood and heart have both improved in character. Arterial thromboses, which are often the result of embolism, though sometimes purely autochthonous, are much more serious, and are apt to lead to senile gangrene of the part to which the artery affected is distributed.

Thrombosis of the gastric veins after middle life is followed by similar results to those that happen at an earlier age; there is pain after food, ulceration, and often severe hæmatemesis. *Results of gastric thrombosis.* Should the ulcer be at or near the pylorus, there may be no vomiting, only dark-coloured stools — melæna.

It is not alone, or chiefly, the altered constitution of the blood that gives rise to the formation of thrombi. It is not even stagnation of the circulation that *Cause of the formation of thrombi.* causes the blood to coagulate in the veins. These may assist, but so long as the endothelium is intact and performs its functions normally, the blood remains fluid.[1] The conclusion from this is, that the sluggish venous circulation has so impaired the vitality of the endothelium that thrombosis is at once precipitated, more especially in certain positions, by whatever further impairs the constitution of the blood or makes its movement more sluggish.

In advanced arterial atherosis, so common in gouty patients, the endothelium occa*Source of gouty emboli.* sionally dies, gets washed off, and so permits calcareous spiculæ to project naked into the vascular lumen; upon which the blood coagulates. In arteries of a moderate size these coagula often become autochthonous thrombi and completely block the artery. In larger vessels the coagula projecting into the blood current occasionally get broken off and carried as emboli into smaller arteries, which they either block completely, or they may block it incompletely, and thus form a nucleus for a thrombus, which ultimately completes the occlusion. In this way embolism of the brain occasionally gives rise to ingravescent symptoms, simulating those caused by recurrent hemorrhages.

[1] *Vide* Cohnheim, *op. cit.*, Vol. i., pp. 174 and 177; also Baumgarten, *Die Sogennante Organisation des Thrombus*, Leipzig, 1877; also Senftleben, Virchow's *Archiv*, lxxvii., S. 421; and Birk, *Das Fibrinferment im lebenden Organismus*, Dorpat, 1880.

Irregularity of nutrition being an indication of the gouty diathesis, it has come to pass that longitudinally ridged (striated) nails have been regarded as a sign of gout (Fig. 8), and so they doubtless are. *Ridged and furrowed nails.* These ridges are not often seen before middle life; they sometimes implicate the whole surface of the nail; at other times there is but one strongly marked ridge from matrix to tip, and even this may be irregularly interrupted by narrow, transverse furrows. These ridges are a sign of the gouty diathesis, but have no connection whatever with a gouty paroxysm; when the nails are thin they often split, and are sometimes very troublesome, but this chiefly in advanced life. Except when they split such nails are more curious than important. It is otherwise with the transverse furrows occasionally found running across the nails (Fig. 9). These are best marked on the thumb-nail, which is the thickest, and are always an indication of a serious illness overlived. Beau in France, and Wilkins here, get the credit of having first directed professional attention to these furrows; but, indeed, their presence and

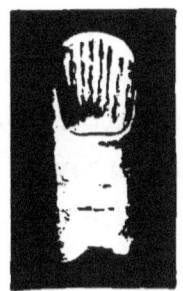

Fig. 8.

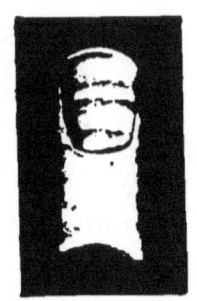

Fig. 9.

signification have been known from time immemorial, and there is not a farrier or horse-cowper who does not understand the importance of a transverse furrow on the hoof of a horse, or who is not quite up to the advantage of new shoes and fresh rasping of the hoof where such a tell-tale exists. As the thumb-nail takes six months to grow from matrix to tip, the position on the nail indicates with tolerable exactness the period elapsed since the illness; it is but seldom that an attack of gout is serious enough to produce such a furrow.

Heberden's knobs are very distinctly connected with the gouty diathesis, though they are not always connected with any paroxysm. Heberden himself says: "What are those little hard knobs, about the size of a small pea, which are frequently seen upon the fingers, particularly a little below the top near the joint? They have no connection with gout, being found in persons who never had it; they continue through life; and being hardly ever attended with pain, or disposed to become sores, are rather unsightly than inconvenient, though they must be some little hindrance to the free use of the fingers." [1] These knobs are common

Heberden's knobs.

---

[1] *Commentaries on the History and Cure of Diseases*, by William Heberden, London, 1803, 2d ed., p. 148.

enough, and there are few physicians of any experience who have not had an opportunity of watching their development. They sometimes

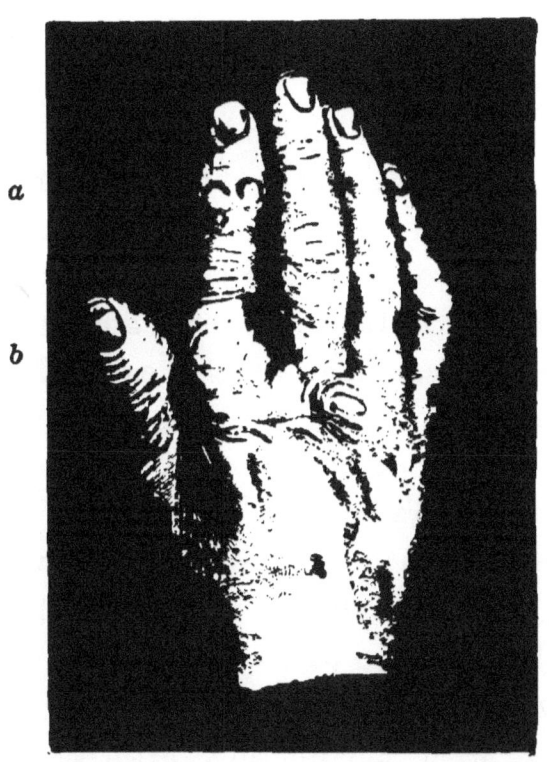

Fig. 10. — At *a* two of Heberden's knobs are seen at the base of the distal phalanx of the forefinger; over the head of its metacarpal bone, *b*, there is a tophaceous mass.

rapidly grow after an acute affection of the fingers, with many of the characteristics of a true gouty paroxysm; at other times they are of slow and gradual growth, accompanied by the ordinary phenomena of gouty dyspepsia, but attended by no more remarkable local phenomena than occasional twinges of pain about the joints, and an occasional sense of fulness and stiffness of the fingers, much aggravated, if not entirely induced, by gastric disturbance. For diagnosis, however, and certainly for treatment, we have to distinguish between Heberden's *knobs* and Haygarth's *nodosities*.[1] The *knobs* are extravascular deposits in the neighbourhood of the smaller joints, chiefly of the fingers, but they may be found about the toes also, and appear as gouty pearls on the cartilage of the ear. They begin like small peas, or at least are scarcely noticed till they are about this size, but they sometimes attain a considerable size, and produce great and irregular deformity of the hands or other parts affected; they are composed of urate of soda, and are popularly known as chalkstones.

The *nodosities*, on the other hand, are associated with rheumatoid arthritis, and not with gout; they are really "exostotic growths, from the margins of the articular surfaces, as well as from the periosteum and bone in the neighbourhood of the diseased joints."[2] These nodosities lead ultimately to

Haygarth's nodosities.

[1] *A Clinical History of Diseases, part 2, of Nodosity of the Joints*, by John Haygarth, Bath, 1805.
[2] *A Treatise on Rheumatic Gout*, by Robert Adams, M.D., London, 1873, p. 16.

anchylosis of the joints; and the deformity of the parts affected, when the hand is at fault, is "invariably associated with a characteristic adduction or inclination of all the fingers towards the ulnar side of the hand."[1] The knobs are due to impurity of the blood, the nodosities to disease of the bone. Gout is peculiar to man; rheumatoid arthritis he shares with the lower animals, notably with the horse. For nearly 100 years the impurity in gouty blood has been known to be uric acid, usually present as urate of soda.[2] This uric acid is due to defective oxidation of the effete material in the blood;

Fig. 11. — Haygarth's nodosities.

[1] Adams, *op. cit.*, p. 252.

[2] In 1797 Tennant and Wollaston established the fact that tophi are composed of urate of soda; but Charcot rightly says, "the period of positive knowledge dates from Garrod's researches in 1848." *Clinical Lectures on Senile and Chronic Diseases*, by J. M. Charcot, New Sydenham Society's Translation, 1881, p. 127. *Vide* also *A Treatise on Gout and Rheumatic Gout*, by Alfred Baring Garrod, M.D., London, 1876, 3d edition, p. 49, etc.

instead of urea being formed and excreted, the lower compound, uric acid, is formed and retained. At certain parts of the body — about the joints, cartilages, and tendons — the circulation, never very active, gets delayed as age advances. The blood-plasma, flooding the tissue interspaces, is reabsorbed but slowly; the urate of soda, never very soluble, crystallizes out on some slight provocation, and gradually grows to gouty pearls on the ear, to Heberden's knobs on the fingers, and to so-called tophaceous deposits elsewhere. The synovial oil lubricating the joints and tendons is less perfectly elaborated than formerly; hence the gouty stiffness and pain on movement, aggravated by a certain amount of tension in the tissue interspaces, which is always present, and is worse at times. Moreover, the uric acid, or urate of soda, not only forms knobs and pearls in the situations specified, but now and then crystallizes within the sheaths of the tendons, notably that of the *tendo Achillis*, giving rise to a grating sensation on movement, which is often painful. The same thing may, indeed, be found in any of the extravascular spaces; for senile remora and the gouty diathesis modify the circulation throughout the whole body, as well as every vital process, whether it be normal or abnormal.

There is one other symptom of gout which deserves special mention, and that is a "peculiar *aura* or rapid twittering motion under the skin, as it were, chiefly in the back and limbs."[1] This twittering of the superficial muscles is limited to a small area; it comes on suddenly and is of short duration; in character it resembles very much an attack of *tremor cordis*, but being quite superficial, it is naturally much less alarming. Repeated attacks of this twittering sometimes precede an attack of gout, but this symptom is often found where only the diathesis prevails. Its causation is quite as inexplicable as that of the *tremor cordis* itself. *Gouty twittering of superficial muscles.*

The infarction theory of what is called a fit of gout, while it accords with and explains all the obvious facts connected with an attack, still leaves many of the more recondite phenomena unexplained. For example: the marked hereditary character of true gout, and the remarkable fact that while all of us acquire the gouty diathesis as age advances, not a tithe of us ever suffer from a paroxysm. Yet these are not more inexplicable than the fact that those who suffer from Heberden's knobs rarely have any so-called fit *Sufferers from Heberden's knobs rarely have a paroxysmal attack.*

[1] *Vide Contributions to Practical Medicine*, by James Begbie, M.D., Edinburgh, 1862, p. 6.

of gout — so rarely that Heberden himself says of these knobs, "they have no connection with gout." Yet these *digitorum nodi* are certainly inseparably connected with the gouty diathesis, of which they are signs as easily recognized and as distinctive as enlarged cervical glands and irregular cicatrices are of struma.

The heredity may be partly of structure; that as yet we do not know. It certainly is of function, and the function is that of the stomach. We know this to be inherited, because long before there can be any question of acquirement we find the gouty dyspepsia in full swing; nay, more, in quite young children we not only find the gouty intolerance of certain articles of food, but we also find that when these articles are consumed their ingestion is followed not only by all the usual dyspeptic symptoms, but also by stiffness and swelling of the digits. Gouty dyspepsia means a feeble and imperfect digestion; occurring in early life, it must be largely a matter of inheritance, though it may be aggravated by injudicious feeding in infancy, and possibly enough it may even be to some extent acquired in this way. I need not say that as care and diet can do much to relieve gouty dyspepsia at any age, so, at an early age, when as yet unaccompanied by any structural alterations,

Gouty dyspepsia is often inherited.

Gouty dyspepsia may be curable.

it may not only be greatly relieved, but may even be cured.

Gouty dyspepsia in advanced life can always be greatly relieved; but as the cause is structural and permanent, watchful care is always a lifelong necessity. The essential element of gouty dyspepsia is feebleness of digestion. The gastric juice, like all the other secretions, is secreted at a low pressure, it is poor in quality, and defective in quantity; hence imperfect digestion. Some of the food escapes the action of the gastric juice, and instead of being formed into healthy chyme it breaks up, under the influence of heat and moisture, into various compounds productive of discomfort in the stomach and of sundry ill effects when absorbed into the blood. The food may undergo acid fermentation, acetic and butyric acids being set free, which irritate the gastric mucous membrane, inducing a catarrhal condition with excess of mucus, which hampers the primary digestion in the stomach, and by extending along the duodenum and bile ducts may interfere with the free passage of the bile and thus impede secondary digestion. A sense of fulness and oppression, with pain and acidity or more often flatulence after meals, indicate that digestion is being interfered with, and result in the fluttering and irregular heart, so usual a concomitant of gouty dyspepsia. The irritated and congested condition of the gastric mucous membrane is the great cause of the gouty *Bulimia*, which is not only the result, but also a very efficient cause, of much of this dyspepsia. Then we have the disturbed sleep — the gouty insomnia, the irregular bowels, and the lateritious sediment in the urine, which together make up those external indications that reveal to the most unobservant the existence of the gouty diathesis.

CHAPTER VIII

THE SENILE HEART, ITS CONCOMITANTS AND SEQUELÆ. GLYCOSURIA, GOUTY KIDNEYS

THERE are several important organs in the body which are very gravely affected by the changes in the circulation due to advancing age, notably the liver and the kidneys. In their turn, the alterations in the structure and functions of these organs, thereby induced, very materially modify all the organic processes during the future progress of life.

Oliver Wendell Holmes says that the most satisfactory and comforting opinion that can be given to a patient, is to tell him that he suffers from congestion of the portal system. And there may be truth in this view; but to tell him that he suffers from too much or too little bile, or from biliousness generally, expressions never out of the mouth of a valetudinarian, is to make use of words without meaning, now that we know that bile is only the

Bile but the drainage of a large manufactory.

drainage of a large manufactory, and is itself apparently of but little use in the animal economy.[1] Indeed, the discomfort which we know to accompany the absence of bile from the stools must nowadays be looked upon as entirely due to the cessation of the manufacture, and not to the absence of the product from the intestinal contents. The amount of bile secreted in a day amounts to somewhat over a pint (638 ccm.),[2] and though this is nearly all reabsorbed, and the movement of the bile and pancreatic secretion may thus be regarded as the analogue of the abdominal circulation of the Gasteropods, yet as these secretions are most

---

[1] "From June, 1890, to the present date, March, 1892, every drop of bile has been poured out on the surface, and there has been no evidence that any has entered the duodenum. Nevertheless, her health and strength have steadily improved. . . . She is fat, and must weigh eleven or twelve stones. She tells me that since her return home she has never had a day's illness, that she is up every morning at her household duties at five o'clock. She states that her appetite is very good, and that she can eat all kinds of food, even the most fatty, with perfect impunity. Her bowels move once a day without medicine. It would be impossible to adduce stronger evidence against the view that bile plays any important part in the digestive process." *Vide* "Further Observations on the Composition and Flow of the Bile in Man." By D. Noël Paton, M.D., *Laboratory Reports of the Royal College of Physicians*, Edinburgh, 1892, Vol. iv., p. 44.

[2] *Vide* "On the Composition, Flow, and Physiological Action of the Bile in Man." By D. Noël Paton, M.D., F.R.C.P. Ed., and John M. Balfour, M.B., C.M., *Laboratory Reports of the Royal College of Physicians*, Edinburgh, 1891, Vol. iii., p. 193.

copious just after the ingestion of a meal, they must to some extent relieve the vascular turgor always greatest at that time. An active liver is a great relief in cases of weak, dilated hearts, and the abdominal circulation just referred to affords a reasonable explanation of this. *Free secretion of bile relieves a weak heart.* In weakly subjects, to obtain this relief it is enough to employ an appropriate cholagogue in a dose sufficient to act upon the liver alone, without purging.

The two great manufactures of the liver are urea and glucose. Urea is the chief ultimate product of the oxidation of nitrogenous bodies, and when these are in excess, or when there is a hypo-oxygenated venosity of the blood, as happens in all more or less after middle life, but especially when the heart gets dilated, then we have the less oxidized product — uric acid — formed, and its neutral salts saturating the system — the gouty diathesis in full swing. Under these conditions there is always congestion, often enlargement of the liver. There is never any difficulty in detecting in the urine the deficiency of urea and the excess of uric acid and its salts; but there is more, for in all cases of congested and gouty liver we get in the urine, with Moore's test (liquor potassæ), a yellow colour which deepens with the congestion, until in many cases we have gouty glycosuria fully developed. There seems to be a regular gradation from the faintest tinge of colour to unmistakable sugar, detectable by every known test, so that it seems a little difficult and somewhat invidious, to say up to this point there has been no sugar, now there is. Mucin (*Nucleoalbumin*) in the urine strikes a yellow colour when the fluid is boiled with liquor potassæ,[1] while uric[2] and glycuronic[3] acids, kreatin and kreatinin,[4] all decompose the copper in Trommer's and in Fehling's tests when boiled with them; and as these are all present in gouty urines, a yellow colour is continually to be found in such urines when these tests are employed. The fermentation test itself may be fallacious, because other matters besides sugar are decomposed under the influence of ferment.[5] So long as the sugar is in a minute quantity, it seems scarcely possible to say whether it is actually present or not; when it is found in a larger amount, the difficulty lies in determining whether we have to do with a true diabetes or

[1] v. Jaksch, *Klinische Diagnostik*, 3e Auflage, Wien u. Leipzig, 1892, S. 327.

[2] v. Jaksch, *op. cit.*, S. 328.

[3] Ashdown, *Proceedings of the Royal Society of Edinburgh*, Vol. xvii., p. 58.

[4] v. Jaksch, *loc. cit.*

[5] Thudichum, *Pathology of the Urine*, London, 1877, p. 429; v. Jaksch, *op. cit.*, S. 329.

merely with a gouty glycosuria. To determine this, we have to fall back upon other subsidiary symptoms.

Gouty glycosuria has a knack of turning up at odd times and in an unexpected man-ner. More than a dozen years ago an elderly gentleman presented himself to me with a dilated heart, an enlarged liver, very considerable general dropsy, marked œdema of the lungs, and about one-third of albumin in his urine, which was scanty. He was puffy all over from general œdema, but seemed also to be well nourished, and had no particular thirst, nor any ravenous appetite; just about the kind of case in which one would least think of looking for sugar in the urine, yet, on examination, over five per cent of glucose was detected. The coexistence of albumin with glucose in the urine is not usually regarded as favourable to the patient, but the prognosis depends, not upon the coexistence of these substances, but upon the probable cause of the presence of them both. In this case the dilated heart was the evident cause; venous congestion of the kidneys leading to albuminuria, and venous congestion of the liver to glycosuria. A dilated heart is an improvable, if not always a curable, organ, even though the dilatation is senile in character, and the old gentleman made a most

Cases of gouty glycosuria.

excellent recovery. His heart improved in a remarkable manner, the dropsy passed entirely away, and the glycosuria disappeared. He lived for several years, and was able to carry on his business comfortably. He had to do a good deal of travelling and occasionally caught cold, and this invariably broke down his cardiac compensation and brought about a return of all his symptoms, but never to so considerable extent as at first. I saw him occasionally at long intervals for these relapses, but the illness was always taken in time, and there was never more than a trace of either sugar or albumin to be found. At first he was dieted, but not strictly, and more for the sake of his heart than of his glycosuria; there was never any subsequent need for this. He died some years ago from pneumonia.

Another old gentleman, sixty-eight years of age, on his way to Vichy, whither he had been sent by two physicians on account of gouty symptoms, was picked up by a London specialist, who detected a considerable quantity of sugar in his urine. This gentleman was told that his disease had been misunderstood, that he had diabetes and not gout, that he need not go to Vichy, but should return home and follow the regimen prescribed, which, if it did not cure him, would certainly alleviate his symptoms. Unfortunately, the patient's former advisers had not previously tested his urine for sugar, and the scene on the patient's return must be left to the imagination. By and by this patient fell into my hands. I ascertained that he was well nourished, and had not been losing flesh; that he had hard, tortuous arteries, and a failing heart; a sluggish liver, not markedly enlarged; and that his urine was loaded with uric acid, which crystallized out as a copious sediment; that the specific gravity of the urine was 1028, and that it contained about five per cent of sugar. I had no difficulty in telling the patient that his former attendants had undoubtedly erred in not ascertaining the presence or absence of sugar, but that otherwise their opinion was entirely correct, and the presence of glycosuria only confirmed their view, and was of no material importance in the case. Naturally this patient had to be dieted for gout, but not for diabetes, which did not exist. In no long time the arterial degeneration began to affect the brain; the mind, hitherto strong and dogmatic, began to waver and have fancies which the patient could not distinguish from realities, though aware there was a difference which he could neither describe nor account for. By and by there was a marked declension of bodily vigour, but the patient kept well nourished to the last. He died of cerebral hemorrhage several years subsequent to the episode referred to, the glycosuria having, to my knowledge, persisted up to a few weeks before his death.

Again, about fourteen years ago a publican chanced to be in a railway accident. Some months subsequently he was found to be passing sugar, and his ailment was dubbed traumatic diabetes by his medical advisers. He was also supposed to be suffering from several obscure nervous ailments due to spinal concussion. In the course of his action against the railway company, his diabetes was represented as of a most serious character, traumatic in origin and due to the accident; six years were assigned as the utmost limit of his life. I found this patient to have a dilated heart and a large liver; also that, like many other publicans, he was a free liver. He was fat and well nourished; and in spite of having passed a considerable quantity of sugar daily for many months, and probably for years, there was not only no emaciation, but also neither thirst nor ravenous appetite. For these and other reasons I had no difficulty in declaring that this patient did not suffer from diabetes, either idiopathic or traumatic; that he had only gouty glycosuria, which in itself would not shorten his days, and which had probably existed for several years before the date of the accident. This patient still survives to attest the correctness of my opinion, in excellent health, passing as much sugar as ever, and in the full enjoyment of the exceptionally heavy damages which the jury awarded him.

These three cases may serve to give an idea of the various circumstances in which gouty glycosuria may turn up, and in which it is of consequence to remember that glycosuria is not diabetes, that the mere presence of sugar in the urine is not a disease, that it is not of uncommon occurrence in gouty people, and that it is specially apt to be found when the heart is dilated and the liver enlarged. In these cases of glycosuria, even when the quantity of sugar passed is considerable (as much as five per cent), there is no emaciation, and there is a possibility of a cure. Strict dieting is quite unnecessary in such cases, as even though the sugar may never disappear from the urine, its persistence is not accompanied by wasting of the body or by any other serious symptom. The sugar seems to be excreted simply because it is in excess of the requirements of the system, either as a result of the superfluity of nutriment ingested, or of a diminished consumption from deficient muscular exertion; probably both of these circumstances have each a share in bringing about the ultimate result. Gouty glycosuria as a rule is easily controlled by regulation of the diet, and many reputed cures of diabetes have probably been cases of this character.

Possibility of a reciprocal action between the kidneys and the heart. The connection between the kidneys and the heart has for long been a subject of great interest to the profession. We know that heart failure gives rise to albuminuria through the intermediacy of congestion of the kidneys, but whether disease of the kidneys is of equal importance in influencing the condition of the heart has long been a subject of controversy, and the literature bearing upon this problem is both voluminous and important.

There is no form of kidney affection, any more than there is any other kind of disease, which is not occasionally associated with disease of the heart in some one or other of its forms, and that either accidentally or for sundry efficient reasons. But there is one form of heart affection so invariably associated with one particular form of kidney disease that, for sixty years past, the relationship has been assumed to be one of cause and effect, while professional opinion has not yet decided which of the two ought to be regarded as the cause and which as the effect.

The almost invariable coincidence of the red contracting kidney with hypertrophy of the left ventricle of the heart did not escape the accurate observation of Bright. He sought an explanation of this in the supposition that "the altered quality of the blood so affects the minute and capillary circulation as to render greater action [of the left ventricle] necessary to force the blood through the distant subdivisions of the vascular system."[1] *The cirrhotic kidney always associated with hypertrophy of the left ventricle of the heart.*

Bright's explanation of this.

But Bright overlooked the fact that in other renal diseases where the blood is also notoriously impure, from the admixture of urinary constituents, no such hypertrophy of the left ventricle occurs. Bright also failed to show that the blood is always impure before the cardiac hypertrophy commences. *Its insufficiency.*

Traube, on the other hand, pointed out that hypertrophy of the left ventricle is not the result of blood impurity, because it does not accompany every form of diffuse renal disease, but is only found in connection with the cirrhotic kidney. And he propounded the doctrine that this hypertrophy is the result of the call for increased exertion made *Traube's view.*

[1] *Guy's Hospital Reports*, Vol. i., p. 396.

upon the heart by the rise of the intra-arterial blood pressure, a rise which he believed to be due to the obliteration of so many arterial branches within the kidneys with the Malpighian tufts attached to them.[1] But Traube overlooked the fact that cardiac hypertrophy is not found associated with every form of contracting kidney, notwithstanding a similar limitation of the intra-renal capillary area, but is only found in connection with the red, granular, cirrhotic kidney.

Also insufficient.

We know, also, that destruction of one or both kidneys, congenital or acquired (cystic kidneys, hydronephrosis, extensive embolic cicatrices, etc.), extirpation of one kidney, amputation of one or more limbs, are all of them accompanied by a much greater limitation of the capillary area than ever happens in any case of cirrhotic kidney, and they are nevertheless entirely without influence in inducing any intra-arterial rise of blood pressure. Traube also neglected to make sure that the heart was not already affected before the commencement of the kidney disease.

According to Sir George Johnson: "The primary and essential structural changes consist in a desquamation, disintegration, and removal of the renal gland-cells, . . . changes in the glandular

---

[1] *Gesammelte Beiträge*, Band ii., S. 290, etc.

epithelium, the result of a modified cell-nutrition, consequent on a morbid condition of blood associated with gout," various forms of dyspepsia, etc.[1] Johnson believes that the kidneys, being no longer able fully to discharge their function, owing to destruction of their tissue, the renal arterioles take on a stop-cock action to cut off that excess of blood which is no longer required, because it can no longer be depurated. This persistent action of the arterioles he naturally believes to result in hypertrophy of their muscular coat. In consequence of the failure of the kidneys to discharge their function the blood is necessarily impure, and more or less unfit for the perfect metabolism of the tissue. To cut off this unsuitable nutriment from the tissues, Johnson supposes that the systemic arterioles also take on a stop-cock action, while for the very necessary purpose of maintaining the circulation at its norm, in spite of this increased peripheral obstruction and consequent rise of blood pressure, the left ventricle is forced to make extra exertion, and consequently hypertrophies.[2] But Johnson's theory postulates too many problems as yet unsolved

Sir George Johnson's pathology of the cirrhotic kidney.

---

[1] Johnson, *Medical Lectures and Essays*, London, 1887, p. 700, etc.

[2] Johnson, *op. cit.*, p. 705.

and unaccepted by modern physiology to be of any pathological value. This theory is also quite incompatible with any relative cardiac hypertrophy preceding the kidney disease. Johnson consequently ignores this.

Is unsatisfactory, because based upon views unaccepted by modern physiology.

Next to Traube's, the theoretic pathology of the cirrhotic kidney which has most impressed the profession, has been that of Gull and Sutton. They deny any direct causal connection between renal cirrhosis and hypertrophy of the left ventricle of the heart. They hold that these two conditions are the result of one general affection of the arterial system, to which they have given the name of arterio-capillary fibrosis, or hyalin-fibroid disease of the arteries.[1] Gull and Sutton acknowledge two forms of contracting kidney. One of these occurs as a local disease, and in most cases, if not in all, is the product of an acute nephritis. This form may occur at any age, and is unattended by any change in the heart, and by very little, if any, recognizable change in any of the other organs in the body. The other form of contracting kidney is not common before forty years of age, is often associated with hypertrophy of the heart, diseased vessels, and more or less widespread changes in

Gull and Sutton's pathology of the cirrhotic kidney.

---

[1] *Medico-Chirurgical Transactions*, 1872, Vol. lv., p. 273, etc.

other organs.[1] In these cases the kidney affection is not always the primary disease, nor can the other organic failures be attributed to the kidney disease. According to Gull and Sutton, the arterio-capillary fibrosis primarily affects the vascular system — to wit, the arterioles; and it invades the other organs — the heart, the kidneys, the lungs, the brain, the spinal cord, etc., not simultaneously, nor in any sequential manner, but as it were casually, as part of a widespread cachexia which has its basis in the vascular system.

But a widespread arterial degeneration, rarely found before the age of forty, has a suspiciously close resemblance to senile degeneration, and the results described as following hyalin-fibroid alteration of the arterial coats are precisely similar to those originating in failure of arterial elasticity. The heart found connected with the cirrhotic kidney is always in the state of dilated hypertrophy usual in the senile heart, varying in degree in each individual case. Primarily this affects the left ventricle, but it is not restricted to it, and it will be found affecting both ventricles, more or less, in every case, and not merely in a matter

This pathology closely resembles, in its history and results, senile degeneration of the ordinary type.

The heart is simply a senile heart.

---

[1] *Lectures on Pathology*, by the late H. G. Sutton, M.B., F.R.C.P., London, 1891, p. 431, etc.

of 70 per cent, as Buhl would have it.[1] This dilated hypertrophy of the heart is also always associated with loss of elasticity and dilatation of the large arteries — the aorta above the valves averaging in circumference 7.6 cm. as against a normal of 6.3 cm.[2] Lastly, this dilated hypertrophy of the heart, as Bamberger,[3] Schroetter,[4] and others tell us, always precedes the kidney affection. Among the scores of senile hearts which have come under my own observation, there have been many with cirrhotic kidneys. In those that proved fatal at a comparatively early stage the heart affection has always seemed to be in advance of the kidney, and in some few I have satisfied myself that this was actually the case.

The cirrhotic kidney, as every one knows, may be contracted to one-half or one-third of its natural size; it is shrivelled, its capsule thickened and opaque, and its surface granular. On section the shrivelling is found to be chiefly at the expense of the cortical portion, and the cut surface is flecked

---

[1] Buhl, *Mittheilungen aus dem pathologischen Institut zu München*, Stuttgart, 1878, S. 64, etc.

[2] This fact is noted by many authors. These figures are taken from Ewald, in *Virchow's Archiv.*, Bd. lxxi., S. 477.

[3] *Lehrbuch der Krankheiten des Herzens*, Wein, 1857, S. 328.

[4] *Ziemssen's Cyclopedia of Practical Medicine*, Vol. vi., p. 192.

with streaks and specks of white from salts of uric acid scattered throughout the cortex and between the tubules. The presence of these salts in the stroma of the kidney has the same significance as elsewhere (*vide antea*, p. 168). It indicates a remora of the circulation sufficient to permit those comparatively insoluble salts to crystallize out of the lymph which floods the tissue interspaces. This remora is due to venous congestion, the result of commencing failure of the central organ of the circulation, and is accompanied by all the usual consequences of venous hyperæmia. One of the consequences of venous congestion of any organ, laid down by Sir William Jenner as a pathological law, is that the structure of any organ so congested becomes hard, tough in texture, and increased in bulk by an exudation of lymph, which is ultimately converted into fibrous tissue. By and by this new-formed tissue contracts, and if the organ be a kidney, its surface becomes uneven and granular, and cysts are developed.[1] The structure of the kidney easily lends itself to these changes. The cysts are readily accounted for by

The cirrhotic kidney a true gouty kidney.

---

[1] *Vide Medico-Chirurgical Transactions*, Vol. xliii., p. 199; and Dickinson's *Diseases of the Kidney*, etc., Part ii., p. 385, etc. *Vide*, also, Schmaus and Horn, *Ueber den Ausgang der cyanotischen Induration der Niere in Granularatrophie*, Wiesbaden, 1893.

the blocking off of some part of a tubule, either at its commencement in a Malpighian capsule, or in some other part of its course; while the pyramids of Ferrein, even in health, present a somewhat granular appearance on the surface of the kidney, and when the septa between them are hypertrophied and contracted, they must largely contribute to the well-known granular aspect of the cirrhotic kidney.

Indeed, when we consider the contractions and deformities that disfigure the comparatively rigid tissues of the extremities of those suffering from the influence of the gouty diathesis, from the vascular changes upon which this diathesis is based, and from those which spring from it, we cannot wonder at the remarkable changes wrought in the softer and more yielding tissues of the kidney by the same means. Hypertrophy and subsequent contraction of the intra-renal fibrous tissue are probably sufficient of themselves to account for all the deformity produced, but the action of these causes cannot fail to be promoted by thrombosis of the vessels, which is of such frequent occurrence in all cases of gouty remora.

Gull and Sutton say: " The morbid state under discussion (arterio-capillary fibrosis) is allied with the conditions of old age, and its area may be said hypothetically to correspond with the 'area vasculosa.'"[1] A statement sufficiently confirmative of all I have been suggesting; but I go a little further, and say that during life it is impossible, even if it were desirable, to differentiate the one condition from the other, and that senile loss of arterial elasticity is sufficient to account for all those sequential changes, which, when excessive, terminate in the gouty or cirrhotic kidney. *Arterio-capillary fibrosis cannot be differentiated during life, if at all, from senile loss of arterial elasticity.*

Atherosis is merely one of those senile changes by which a structured matrix is replaced by amorphous material (*vide antea*, p. 13). Gout, and such poisons as alcohol, lead, and syphilis, promote the advent of this change, and the last-named poison is specially responsible for the end-arteritis so often present. But these special conditions do not seem to be necessary for the production of the gouty kidney, though undoubtedly they precipitate and intensify all those sequential changes which find their natural termination in this structural alteration.

Sir George Johnson's uncontradicted statement, that the cirrhotic kidney is " of common occurrence in those that eat and drink to excess,"[2]

---

[1] *Medico-Chirurgical Transactions*, Vol. lv., p. 296.
[2] *Medical Lectures and Essays*, London, 1887, p. 680.

sufficiently explains the predominance of hypertrophy in the hearts of such patients; while the fact that the cirrhotic kidney is not restricted to gross feeders, but is also found in those labouring under " certain forms of dyspepsia,"[1] accounts for this affection not being restricted to those with marked cardiac hypertrophy. Every case requires to be considered and explained by its own individual circumstances.

It must be remembered that dilatation is the first stage of senile cardiac failure (*vide antea*, p. 33), and even when hypertrophy afterwards becomes excessive, the ventricular cavity will always be found increased in size, though in some cases the mode of death makes the *post-mortem* appearance of the heart apparently to belie this (so-called concentric hypertrophy).

We are told that it is always easy to differentiate the congested kidney of cardiac failure from the cirrhotic kidney; because in the former case the urine is diminished in quantity, and there is a considerable amount of albumin present; while in the latter case the urine is not diminished, often greatly increased, and the albumin is present in but small quantity, often only as a mere trace. This statement is perfectly correct, and yet in my own experience some indication of cardiac failure

[1] Johnson, *op. cit.*, p. 680.

has always preceded any manifestation of kidney affection.

In those rare instances in which a lifelong acquaintance with all the details of the case has made known every point in its history, there is no difficulty in ascertaining this with at least the highest probability.

When, however, a case of senile heart, seen for the first time, presents a trace of albumin in urine otherwise normal as to quantity and quality, without any evident soakage of the tissues, we are not, perhaps, justified in regarding the kidneys as cirrhotic; but we may be well assured that only careful treatment can postpone, or possibly prevent, such an untoward ending.

Having been for nearly all my professional life fully cognizant of the medical history of a well-known dignitary of a northern university, and having been for many years his occasional medical adviser, I was well aware that he had hard and tortuous arteries and a hypertrophied heart, and had my eyes fully open to all the contingencies likely to happen in such a case.

Case of senile heart with probable cirrhotic kidneys. Fatal by uræmic coma.

For years, however, the patient went on the even tenour of his way. At last, after passing his grand climacteric, his heart gave way: it became dilated, with a systolic bruit in the mitral area. But not for some time after this, not till about four months only before the patient's death, could any albumin be detected in his urine. This albumin never amounted to more than one-third, often to much less, varying from time to time, and, so far as I know, it was never afterwards absent till the end.

The albumin came with failing health and a broken constitution; the outward frame looked vigorous still, but the organization was giving way at all points, and revealing its failure in many ways. To an ordinary observer, it seemed as if the patient would be at once restored to his pristine vigour, if the mysterious disorder that sapped his strength could be recognized and removed. To the intelligent eye of the physician, there was but one possible ending, though it might come in various ways, and might be warded off for an uncertain period by careful and judicious treatment.

The hard, tortuous arteries, the dilated heart, and the albuminous urine, told an unmistakable tale to a discerning mind, and in no long time this was emphasized by a sudden attack of blindness of one eye, due to hemorrhage into the retina from rupture of a vessel, as ascertained on examination by one of our ablest oculists.

The patient had almost constant headache, a feeling of intense depression, and a sensation of failure, pathetically revealed by a longing for a quiet life, and release from the burden of official duties; a quietness which he scorned, and a burden which was unfelt in the plenitude of health, for in those days there were few who could compare with him in fulness of life and energy. Part of his depression was doubtless constitutional, as he had suffered similarly at an earlier period of life, but at this time there was every reason to believe that this constitutional depression was intensified by imperfect nutrition of the brain, due to arterial atherosis. Of this condition the hard external arteries, and especially the retinal hemorrhage, must be accepted as a sufficient exponent.

Already the outworks were sapped, and the enemy was marching along the pathway of the arteries towards three breaches in the enceinte of the citadel of life — the dilated heart, the contracting kidneys, and the shrinking brain. Which of these would be the first to be stormed it was impossible to foretell.

There was nothing to be done but to send the poor patient to a milder climate for the winter, in the hope that his valuable life might be prolonged. A full and particular account of his illness and present condition accompanied him south, and the programme indicated was carried out. About a month after I last saw him the patient died somewhat suddenly from uræmic coma.

This case was in every respect one of intense interest. What is specially noteworthy is, that there is an absolute certainty that the hypertrophy of the heart long preceded any indication of kidney affection, though this was carefully looked for during many years; that the kidney affection never became cognizable till the heart began to fail; and, lastly, that this cardiac failure was quite unaccompanied by any other sign or symptom, apart from those of the heart itself, beyond the slight albuminuria; in particular, there never was the slightest trace of dropsy, nor any detectable soakage of the tissues.

Case of probable cirrhotic kidney, fatal from hepatic hemorrhage.

The following case also inculcates the same lesson: that even when hypertrophy is the predominant lesion of the heart the kidney affection only becomes cognizable when this organ begins to fail.

For nearly forty years I was intimately acquainted with a gentleman, who ultimately died at the age of sixty-eight, and I had been his medical attendant for some considerable portion of that time. He was a man of robust frame, who lived well, and enjoyed excellent health up to a couple of years before his death.

Early in life this patient's arteries became large, hard, and tortuous, and for quite ten years before his death he was known to have a large hypertrophied heart, which had come on insidiously and presented no symptoms. About five years before his death his heart began to falter and become irregular; it had commenced to fail. This irregularity was never altogether remedied, but it did not increase. During all this time the state of the kidneys had been watched most carefully, but not the slightest imperfection could be detected, till about two years before the patient's death, when a small amount of albumin, little more than a trace, was at last discovered. Henceforward a trace of albumin was, with only rare and occasional exceptions, always to be found in the urine. About the same time, two years before his death, this patient began to suffer from defective memory, and to show in other ways a loss of brain power, of which he himself was painfully conscious.

Precisely as in the former case, the enemy had already seized the outworks, and was marching along the arteries upon the citadel of life, in which the same three breaches had been made. In this case it was evident that the brain had suffered most, and it seemed probable that the breaking of the "golden bowl" would have closed the last scene of a most useful and energetic life.

As it happened, this was not the case. For long this patient had suffered from a contracting liver without any marked symptoms beyond constipation and troublesome piles; now, however, jaundice set in, and after a few months of suffering, intestinal hemorrhage suddenly closed the scene.

In this case, also, there was an absolute certainty that cardiac hypertrophy had for many years preceded any manifestation of kidney disease; the kidney affection never became cognizable till the heart began to fail; and, lastly, this cardiac failure was unaccompanied by the slightest trace of dropsy, or of any soakage of the tissues.

The absence of a *post-mortem* examination is a certain disadvantage in founding any argument upon these two cases. I think, however, that both Gull and Sutton would have freely acknowledged both as presenting well-marked clinical symptoms of arterio-capillary fibrosis, while I look upon them both as excellent examples of senile heart originating in senile loss of arterial elasticity.

In all such cases a gouty kidney is one of the possible sequential phenomena, slowly and gradually developing itself out of the slowly increasing venous hyperæmia, but never betraying itself by sign or symptom until, from some cause or other, there is some evident failure of the myocardium.

Even after this the cardiac failure is so slowly progressive that the conditions remain for long practically unchanged, or they may even improve under treatment. But the declension, though gradual, is sure, and the end comes at last, but not always through the kidneys. One advantage of taking this view of the gouty kidney is that the senile variety may be looked upon as preventible, and in the early stages even as amenable to treatment. The gouty kidney occurring before middle life is, however, associated with too pronounced an arterial affection to be treated with success, though even in such a case the symptoms may be ameliorated, and the end postponed by judicious care.

The disadvantage, as well as the advantage, of regarding the cirrhotic kidney as a sequence of senile degeneration.

CHAPTER IX

THE THERAPEUTICS OF THE SENILE HEART. GENERALITIES

THE heart is the one organ of the body whose sufferings are most apt to disturb the equanimity even of the most imperturbable. We know that with each pulsation, life and intelligence are flashed to the farthest outpost of our frame, and we also know that if the heart-beats falter for a second or two we fall to the ground, pale, limp, and almost inanimate — an almost which speedily becomes absolute, if from any cause these heart-beats are prevented from resuming their pristine vigour. With this knowledge ever before our eyes, and clinched by many a startling fact, we cannot wonder that feelings of alarm are excited by any deviation from the normal which makes us cognizant of the movements of so important an organ, of which we are ordinarily so profoundly unconscious. Hence palpitation, intermission, irregu-

Cardiac troubles always alarming.

larity, and *tremor cordis*, all of which make themselves so disagreeably perceptible to our senses, appeal most forcibly to the imagination of the patient, and bring him more certainly to the physician than cardiac ailments of more serious import but of less obtrusive character.

Symptoms such as those described always, and at every age, indicate some physical impairment, a matter of comparatively little moment in early life, but of very much more serious import after middle life. We must not forget, too, that at any age, but more probably in advanced life, the physical impairment may be primarily due to failure of the trophic nerve centres.[1] The marked improvement that ordinarily follows treatment shows that a primary lesion of the nerve centre cannot be of frequent occurrence, though we may accept it as a possible explanation of the intractability of some of the cases we meet with.

Senile diseases are always degenerative, and tend to precipitate the natural termination of life. The

[1] "In the human body there is no mover that can properly be called FIRST, or whose motion does not depend on something else. . . . The contraction of the heart is indeed the cause of the circulation of the blood, and consequently of the secretion of the spirits (as is supposed) in the cerebellum, etc.; but without these spirits this action of the heart could not be performed. These two causes, therefore, truly act in a circle, and may be considered mutually as cause and effect." — Whytt, *On Vital Motions*, Edinburgh, 1751, p. 270.

object of treatment in senile affections is not therefore quite the same as in the diseases of earlier life. We no longer hope for complete restoration, but we expect to alleviate suffering, and to check decadence; and so far as the heart is concerned, we are generally able to attain both of these objects even long after the average limit of mortality is overpassed.

But may be remedied at any age.

Many years ago a gentleman of seventy-seven years of age consulted me as to severe fainting fits to which he was liable. A distinguished consultant, since dead, had told him that these attacks were due to fatty degeneration of his heart, and that no treatment would be of any avail. I found the heart's impulse imperceptible, the sounds faint but pure, the arteries firm, but neither hard nor tortuous; the urine was free from albumin, and of average specific gravity. I explained that, considering his age and the then state of medical opinion, his adviser was perfectly justified in both his diagnosis and prognosis, but I added that experience had taught me that hearts supposed to be fatty, were often only weak,[1] and that so serious a prognosis could only be justified by failure of treatment. The result of treatment in this case was a steady improvement in health and in force of heart-beat, and the patient did not die till he

[1] Balfour, *op. cit.*, p. 309.

attained the ripe old age of ninety, and then not from his heart at all, but from senile asthenia.

In this fight with mortality, medicines have no doubt their place and power, but it is attention to the little things of daily life — the little things of eating, drinking, and doing — that influence the patient's comfort, and gradually turn the scale of health in his favour. Herein lies one of the great difficulties in the way of successful treatment, for when the regulations of the physician come to be pitted against the habits of a lifetime, there is sometimes a difficulty in securing acquiescence.

For several reasons (*vide antea*, p. 34) I have given the senile heart the synonym of the gouty heart, but the connection between the two is more obvious at one time than at another. Thus, some years ago I used to be favoured with occasional visits from an elderly lady with an irritable and slightly dilated heart, which I told her could be best described by the term gouty. Then she used to turn upon me and say, " But Doctor this and Doctor that," for she was a great patron of doctors, " all say what does Dr. Balfour mean by saying you have a gouty heart, for you have no gout." To this my reply was, " So much the worse for you ; if you were to have a fit of gout, your heart would probably be relieved." And doubtless this would have been

Two varieties of irritable heart.

the case, *quoad* the irritability at least. Years afterwards this old lady died, and I ascertained that her irritable heart was due to a weakness for porter in excess.

Again, I often see an old lady, hale and well preserved for her years, which are somewhere over eighty. For long this patient has had the most pronounced gouty heart of its kind I have ever seen. She has never had a regular fit of gout, but her fingers are all distorted with Heberden's knobs; she has constant dyspepsia of a well-marked gouty character, and she has a weak, irritable, and somewhat dilated heart, whose most forcible attempt at a beat is often but a mere flutter for days at a time. Yet she is a sober, careful-living woman, and her heart responds well to treatment even at her advanced period of life. I have never heard a doubt hinted as to the nature of her complaint, though wonder has often been expressed at her apparently marvellous recoveries.

These are opposite extremes, variants of the senile heart in which irritability is the prevalent characteristic. In the one case I wasted much good advice as to what to eat, drink, and avoid, which if attended to would have sufficed to cure the patient. In the other case the long-continued gouty dyspepsia, together with Heberden's knobs on her fingers, were proof enough that something more was required than a merely dietetic treatment.

In our dealings with senile heart affections, we must not forget that all cardiac affections found in the old are not necessarily senile in character, though they must all be unfavourably modified by the conditions present. At this moment I am acquainted with a hale old gentleman, of eighty-six years of age, who for sixty-six of these years is known to have suffered from a dilated and hypertrophied heart. *All heart affections in the old not necessarily senile in character or origin.*

Sixty-six years is certainly the longest period, in my experience, mitral regurgitation has been known or even surmised to exist. But I am well acquainted with many cases in which cardiac disease of various forms, both mitral and aortic regurgitation and more rarely mitral obstruction, has been certainly known to exist from youth to age for periods varying from forty to fifty, or even more years, without any marked discomfort, except when compensation has been temporarily ruptured by illness. Many of these sufferers have led very active lives; some of them have been members of my own profession, who have shirked no work however hard; and it has seemed to me that the most active have suffered least. Possibly because *Long duration of some cardiac affections.*

the disease was not so serious, certainly to some extent because the heart is a muscular organ, and like all such organs is strengthened and invigorated by exertion not carried to exhaustion.

Irritability, with more or less of cardiac uneasiness (*vide antea*, p. 35), is one of the earliest indications of advancing senility of the heart. The patient conplains of an uneasy empty feeling in the precordial region; sometimes this uneasy feeling amounts to actual pain in and around the heart, but a pain strictly localized and neither shooting nor darting in any direction. Along with this uneasiness there is irritability of the heart's action, both as to rate and rhythm. There may be palpitation; rapid action; simple intermission at regular or irregular intervals, the heart simply dropping a beat occasionally; or this intermission may continue during periods of longer or shorter duration, and may occur at longer or shorter intervals, and mostly as the result of emotion or of gastric disturbance; or lastly, the heart's action may be more or less persistently irregular as to rate, rhythm, and force simultaneously.

Earliest symptoms of the senile heart.

These phenomena always indicate debility of the myocardium, which, left to itself, sooner or later leads to dilatation of its cavities, after a fashion already explained (*vide antea*, p. 40), with all the serious consequences which flow from this condition.

These sequential events do not follow a similar course in every case, but each follows its own course, according to laws which may be more or less easily recognized.

One patient may for years complain of nothing more than an occasional soreness in the cardiac region, and at last break down suddenly, from what he flatters himself is only neurasthenia, but which turns out to be merely a commonplace dilatation of the heart. This may end slowly by dropsical asthenia in the usual way; not infrequently it terminates in a fatal attack of angina of the ordinary form; or, perhaps even more commonly, in that form of sudden cardiac failure which may be called *angina sine dolore*. Another patient may only complain of occasional intermission, or of fluttering of the heart, — *tremor cordis*, — which annoys him by its recurrence, and such a case terminates perhaps more often by an attack of cardiac syncope — *angina sine dolore* — than in any other way; while there are still others in whom intermission, irregularity, or *tremor cordis* persist for many years without any apparent detriment. In time, however, such symptoms, unless

Modes in which the senile heart may develop.

remedied by treatment, ultimately terminate in serious cardiac disease.

Many such patients seem to suffer but little from their ailment; it seems somehow to escape their cognizance; but there are others, not more seriously ill, who suffer very much from the feeling of insecurity engendered by their malady. "I have not gone to bed for months," said a comparatively young woman to me lately, "without leaving everything as straight as possible. I feared each night would be my last." Yet her only detectable malady was a somewhat marked irregularity of the heart's action, which was completely removed after about one month's treatment. Then her remark to me was of a very different character. "I feel quite well and young again. I had a race down our avenue the other day and felt neither breathlessness nor irregularity."

Cardiac disturbance apt to engender feelings of insecurity.

With the cardiac irregularities and intermissions of the aged there is so often a faltering of consciousness, or a failure of muscular power, that, as a rule, paralysis or brain failure is more dreaded than failure of the heart. Yet there is probably no sufferer from *tremor cordis* who does not feel inclined to exclaim with Sir Walter Scott, "What a detestable feeling this fluttering of the heart is."[1] We cannot even flatter ourselves with Sir Walter that it is confined to the erudite — *morbus eruditorum* he called it — any more than we can nowadays limit with Sydenham the *podagra*, upon which it depends, to the great and noble.[2]

The senile heart is, as we have seen, a term which comprehends many symptoms and a variety of signs, and is essentially a cardiac failure based upon imperfect metabolism. It is therefore of the greatest consequence to determine the cause of this failure by ascertaining the source of the malnutrition upon which it depends.

Senile cardiac failure essentially based upon imperfect metabolism.

---

[1] "I know that it is nothing organic, and that it is entirely nervous, but the sickening effects of it are dispiriting to a degree. Is it the body brings it on the mind, or the mind that inflicts it on the body? I cannot tell; but it is a severe price to pay for the *fata morgana* with which fancy sometimes amuses men of warm imaginations. As to body and mind, I fancy I might as well inquire whether the fiddle or the fiddlestick makes the tune. In youth this complaint used to throw me into involuntary passions of causeless tears. But I will drive it away in the country by exercise." — *Journal*, Vol. i., p. 153.

[2] "Gout kills more rich men than poor, more wise men than simple. Great kings, emperors, generals, admirals, and philosophers have all died of gout. Hereby Nature shows her impartiality, since those whom she favours in one way 'she afflicts in another." — *Works of Thomas Sydenham, M.D.*, translated for the Sydenham Society by R. G. Latham, M.D., Vol. i., p. 129.

In all these cases the objective phenomena are most to be relied upon. The subjective phenomena, the symptoms complained of, are chiefly valuable as corroborating or explaining the information derived from direct observation. Even in regard to what might be regarded as so unmistakable a disease as angina this statement holds good (*vide antea*, p. 86), however trustworthy the patient may be.

In diagnosticating such cases objective phenomena are more to be relied upon than those which are merely subjective.

In examining such a case, with a view to treatment, the pulse is one of those factors which requires careful consideration; rate, rhythm, and especially tension, all being elements of the greatest importance in formulating any opinion as to the exact nature of the case. If the pulse be small, soft, and compressible, the blood pressure is low, and the blood probably deficient in quantity (anæmia), more often defective in quality (spanæmia), as the simple anæmic condition following a hemorrhage speedily becomes spanæmic from the absorption of fluid from the tissues. In such a case we must inquire into every possible source of loss of blood, every possible cause of hæmolysis, and into every conceivable kind of interference with hæmogenesis. External hemorrhages and

Pulse and blood must both be carefully examined.

suppurations are too obvious drains to require more than a mere casual mention. At one time one pretty free hemorrhage starts the organism on a downward career which there is no arresting; at another time an inconspicuous dribble — very often intestinal in origin — slowly and imperceptibly drains the life away. *The indications to be drawn from low blood pressure.* Into these we must not only inquire by interrogation, but we must, so far as possible, scan every part of the mucous tract from the nostrils to anus, and also investigate the nature and frequency of all the natural discharges, for a trifling but persistent diarrhœa may sap the strength and give rise to many anomalous symptoms; or, more insidious still, an excess in venery, trifling and moderate as it may seem, may yet come to be a serious drain when coupled with the impeded hæmogenesis of advancing age. In the early stages of many serious complaints which may start a senile heart of the most serious description a great deal of information is to be obtained from an examination of the blood itself. Thus we have an increase of the white cells in leucocythæmia and Hodgkin's disease; the poikilocytosis of pernicious anæmia; or the mere deficiency of hæmoglobin, in youth suggestive only of chlorosis, but in advanced life hinting at some obscure malignant disease. Failing any discoverable source of hæmolysis, or any gross interference with hæmogenesis, we have always to deal with a certain amount of dyspepsia, which is often a result, and not a cause, of the heart failure, but which always materially influences the composition of the blood, and must, therefore, be considered and provided for in any treatment which is to prove effectual.

If, on the other hand, the pulse is firm, hard, wiry, and full between the beats, this indicates an abnormally high blood pressure. That is the name we give to the condition; the fact that underlies this, and which is really what is indicated, is that *Indications to be drawn from the presence of a high blood pressure.* the blood does not pass so freely as it ought from the arteries into the veins. This condition of pulse may coexist with any of the conditions just referred to, and this additional element in no ways makes unnecessary our enquiry into the state of the blood itself.

The normal loss of arterial elasticity, which accompanies advancing years, of necessity increases peripheral friction, raises the blood pressure, and thus increases the work of the heart (*vide antea*, p. 13). Under the influence of rheumatism, gout, or of such poisons as alcohol, lead, or syphilis, frequently accompanied by acute or chronic inflammation of one or other of the arterial coats, or of all three, the arteries get converted into hard, often ringed and rigid, tubes, whereby their elasticity is still farther diminished, and the work of the heart much increased. Normal nutrition is quite sufficient to maintain the heart intact under ordinary circumstances; but any interference with the nutrition of the heart on the one hand, or any considerable increase of its work on the other, disturbs the equilibrium, and the heart becomes irritable, and it may be irregular in its action, and slowly dilates. On the other hand, if a similar call for exertion on the part of the heart is accompanied by a superabundant supply of food and stimulants, the primary dilatation [1] is speedily followed by considerable hypertrophy of the left ventricle, and we have established the "luxus" heart of the Germans. This *A "luxus" heart by no means idiopathic in its origin.* hypertrophy of gourmands is described by Traube and Fraentzel as due to luxurious feeding alone; but, as Cohnheim has wisely said, "To an accurate comprehension of the manner in which superabundant meals increase the work of the heart, the physiological data at our command are inadequate." [2] There is, in truth, no fact in physiology that teaches us that excess of nutriment promotes cardiac hypertrophy. There is no hypertrophy of

[1] Cohnheim, *op. cit.*, p. 66.   [2] Cohnheim, *op. cit.*, p. 67.

the heart among the Strassburg geese stuffed to repletion to supply the market with *foie gras*, nor did any one ever hear of a young porker, fattened for the butcher, having enlargement of the heart. As an initiative, there must first be peripheral obstruction, so that the heart, being "more than usually exercised in its office,"[1] hypertrophies. If the call for extra exertion is only imperfectly responded to, because the heart is ill nourished, irritability and dilatation of the myocardium speedily follow. But if the obstruction is considerable, and the blood-supply abundantly nutritious and stimulating, hypertrophy follows. Both processes progress slowly and gradually; but the irritability of a weak dilating heart very soon attracts attention to itself; whereas, a hypertrophied heart gives rise to but few symptoms, and is apt to be overlooked and only discovered accidentally, until it begins to fail. A heart in which the left ventricle is hypertrophied, with a hard, firm pulse indicating increased intra-arterial blood pressure, is often looked upon as renal in its origin, because frequently found associated with cirrhotic kidneys (*vide antea*, p. 200). It is, however, much more probably primary, and the cause of the kidney affection, not its result (*vide antea*, p. 202).

[1] *Lectures on Surgical Pathology*, by James Paget, F.R.S., etc., London, 1870, 3d edition, p. 49.

Even in young and healthy arteries any increase of the blood pressure over the normal mean produces rigidity of the arterial walls.[1] In advancing age, therefore, as the arterial walls become less elastic, a comparatively trifling rise of the intra-arterial blood pressure suffices to make the arterial coats tense and rigid, the artery rolling like whipcord beneath the finger, while the trifling character of the pulsatile movement, coupled with the considerable pressure required to arrest it, sufficiently indicate the nature of the obstacle to the ventricular output, and the consequent embarrassment to the heart's action. An embarrassment of this character occurring during the night is a common cause of cardiac asthma, and evidently many deaths from *angina sine dolore*, whether by day or night, are due to this cause, happening as they so often do when the patient is at perfect rest. The spasm of the arterioles, which in such cases is the cause of the rise of the blood pressure, may be reflex from the stomach, intestines, or some other organ; or it may be due to direct irritation of the vaso-motor centre in the medulla. There is even reason to believe that at times the rise of

---

[1] "70 or 80 mm. of mercury is about the normal mean blood pressure of the rabbit, and this experiment shows that above this the arteries become more and more rigid-walled." — Roy and Adami, *Practitioner*, 1890, p. 351.

blood pressure may be due to increase of the ventricular output from cardiac stimulation, and there is no reason to doubt that cardiac stimulation coupled with contraction of the arterioles is a very frequent cause of this rise of the intra-arterial blood pressure. A great effect in this respect is usually accorded to the kidneys; this, however, can only be partially tenable, as it is only universally applicable in the case of the cirrhotic kidney, and in that affection there is sufficient reason for the rise of blood pressure apart from any affection of the kidney, which it indeed precedes (*vide antea*, p. 202).

CHAPTER X

THE THERAPEUTICS OF THE SENILE HEART. EXERCISE AND DIET

WHEN any one past middle life complains of symptoms resembling those described as associated with the senile heart, we know that, whatever else there may be, there certainly is failure of the myocardium. In such a case our first endeavour must be to discover and remove any possible cause of enfeeblement, whether that be an obvious drain or merely a constructive one, nervous exhaustion following overwork or worry. And our next endeavour must be to build up and energize the frame generally, and the heart in particular. Exercise, diet, and medicines are the three agents employed to this end.

Medicines are indispensable in restoring the vital balance of an organism that has been lost through failure of an organ, especially if that organ be the heart.

Exercise and diet are, however, paramount in

maintaining the integrity of a healthy organism; and, properly employed, they are also of the greatest value in restoring it when lost.

Of exercise in the treatment of the senile heart.

It seems somewhat of a paradox to speak of exercise as a treatment for an organ which takes its needful rest in sections, and that only for fractions of a minute, and which, as a whole, is in constant and continuous work from man's birth to his death. Yet we know that, like every other muscular organ, the heart is strengthened by exertion, and that if well fed, it hypertrophies when that is in excess. The rational deduction from this is that exercise judiciously employed may be profitable to the strengthening of a weak heart. Stokes was the first to point out this. He recommended graduated exercise as useful in the treatment of those weak hearts which he believed to be the subjects of fatty degeneration;[1] and from a personal reminiscence of von Ziemssen we learn that Stokes also employed exercise in the treatment of valvular lesions, and specially insisted on the value of even violent exertion in the treatment of aortic regurgitation.[2] Of late years Oertel has done good service in directing the attention of the

---

[1] *Diseases of the Heart and Aorta,* Dublin, 1854, p. 357.

[2] *Verhandlungen des Congresses für Innere Medicin,* Wiesbaden, 1888, S. 55.

profession to the importance of regulated diet and exercise in the treatment of cardiac affections. Much benefit has doubtless followed the recognition of the fact that the discovery of a cardiac murmur is not a signal for a carrying chair, and need not be looked upon as a bar to moderate exertion. Yet, though aortic regurgitation is not always found to be a bar to even violent exertion, no one, I think, would be inclined to treat it, as Stokes is said to have done, by setting the sufferer to run behind his own carriage.[1] Regular, moderate exertion helps to keep the myocardium well nourished, whatever goes beyond tends to promote hypertrophy; and as the coronary arteries have only a limited feeding power, when the myocardium gets beyond this, irremediable failure inaugurates the end under various symptoms. Nor is this the only risk; for long-continued exertion, especially if violent, is, we know, liable to be followed by muscular collapse, and what this means to a heart it is not difficult to imagine. Moreover, scarcely a day passes in which the danger of irregular exertion is not exemplified by the sudden death of some one hurrying to catch a train or a 'bus — a death which often puts a sudden termination to the useful lives of those who had never been known to ail, although they had certainly begun to age.

[1] *Loc. cit.*

When, however, the compensation is only incomplete, when exertion, even though slight, brings on dyspnoea, even if there be no evident soakage of the tissues, and still more if there is, then exercise, even though carefully graduated, ought to form no part of the treatment: the risk is too great. Rest, diet, and heart tonics will suffice, if it be at all possible, in time to restore the compensation, and then exercise may be hopefully resorted to as an adjuvant, but it must be begun cautiously and continued with care. In a great many cases of senile heart, intermission and irregularity are entirely reflex in character, and are not increased, but rather relieved, by exercise, the effect of which is to lower the blood pressure,[1] and thus to promote the fulness and freedom of the heart's contraction, as well as its force and vigour. The exercise not only benefits the heart at the time, but by promoting the circulation through its walls it nourishes the muscle and accumulates energy within the ganglia. This, however, can only happen when the organism is not enfeebled, and when the heart itself retains

Rest is, however, absolutely paramount in certain cases.

---

[1] *Vide* Foster's *Physiology*, fifth edition, p. 148: "At the time of contraction more blood flows through the muscle, and this increased flow continues for some little time after the contraction of the muscle has ceased."

sufficient recuperative power, and is more oppressed than debilitated. The greater freedom of respiration and of circulation resulting from exercise makes metabolism more perfect, and thus favourably influences the manufacture of urea in the liver, and promotes the depuration of the blood. Thus, in appropriate cases, not the heart only, but the whole organism, is the better for exercise.

In most cases of senile heart, however, rest will be found the most generally applicable treatment, and when palpitation, irregularity, or breathlessness follows exertion, rest is the treatment to which — at first, at all events — we are restricted.

The question of exercise must always be carefully considered in relation to each special case, and thereafter adapted and regulated in accordance with its requirements. But, while in regard to the relief of symptoms the question of exercise always requires careful consideration, there is no doubt whatever that as a preventative of many of the evils associated with the senile heart it holds a most important place ; not the foremost place, nor the paramount position, for that belongs to temperance alone, which even without exercise can maintain health, if it cannot bestow strength nor ensure longevity. Of this a late eminent physician was a notable example, and there is not one of us who cannot call to mind many similar instances.

Temperance — moderation in all things — is the true secret for preserving a *mens sana in corpore sano;* and if it be not a certain passport to longevity, it at least enables us to live healthily for as long as we may. In these *fin de siècle* days, when every doctrine is a fad and gets pushed to an extreme, there are multitudes eager to enforce a rabid teetotalism upon all their fellow-men as the only panacea for health, happiness, and longevity. But if we except the votaries of vegetarianism, which is more of a cult than a protest against excess, I know of no society that inculcates, by precept or example, temperance in regard to food; yet there is nothing ages a man or a woman so rapidly, there is nothing that shortens life so certainly, and there is nothing that embitters the latter days of life so much as over-indulgence in food. To those who can afford thus to transgress — to the well-to-do — excess in food is a much more serious menace to health and life than excess in drink, and it is specially so in respect of senile affections of the heart, some of which have been distinctly recognized to owe their origin to over-indulgence, while all are distinctly aggravated by it.

Importance of temperance.

All those who after middle life complain of cardiac symptoms require to be dieted for some reason or other; the condition of the patient and his leading symptoms supply the indications for which we have to provide.

Various dietaries may be required.

The larger number of such patients are either at their normal weight or slightly below it; they often suffer very considerably from intermission or irregularity of the heart's action, with breathlessness on exertion. These require careful regulation of a normal dietary presently to be specified. A smaller number are over their normal weight,— obese,— and suffer more from breathlessness and less from irregularity than the preceding class of cases. These require to be specially dieted and cared for, so as to remove the obesity without diminishing the cardiac energy or the strength of the myocardium. Lastly, we have those in whom there is more evident failure of the myocardium. There may not be so much trouble from intermission and irregularity, but the signs of cardiac dilatation are more marked than in either of the preceding classes of cases, and there are more or less evident indications of soakage of the tissues. Such cases require to have a specially dry diet prescribed for them.

In all cases where diet and dietaries come into

question, the first point of importance is to divide the day properly, so that there may be a sufficient interval between the meals. This is a matter of absolute necessity to secure perfect digestion.

Number of meals and length of interval between them.

Three things greatly disturb gastric comfort, — too large a meal, too short an interval between the meals, and, lastly, the ingestion of food into a stomach still digesting. If the heart is weak, the discomfort induced by such irregularities is, after middle life, more apt to be felt in connection with that organ than in the stomach itself. In health, the stomach empties itself in from three to four hours after the ingestion of a meal, and requires an hour's rest before a further supply is introduced. In those with weak hearts and feeble circulations, the digestion is necessarily somewhat slower; hence the first rule to lay down is: *There must not be less than five hours between each meal.*[1] This allows of three meals in the day, with a sufficient interval after the last meal to permit its digestion to be well advanced before

---

[1] Abercrombie says: "If digestion goes on more slowly and more imperfectly than in the healthy state, another important rule will be, not to take in additional food until time has been given for the solution of the former. If the healthy period be four or five hours, the dyspeptic should probably allow six or seven." — *Pathological and Practical Researches on Diseases of the Stomach*, London, 1837, 3d edition, p. 72.

retiring to rest, which tends to ensure a quiet and restful night. The next matter of importance to remember is, that the ingestion of solid food into a stomach still engaged in digesting a former meal arrests the process and provokes the formation of flatulence; hence the second rule to be laid down is: *No solid food of any kind is to be taken between meals.* This rule is absolute; not a morsel of cake, or of biscuit, or any similar trifle, is to be ingested between meals. There is nothing so destructive of gastric comfort as the continual pecking induced by gouty bulimia. This prohibition does not extend to fluids, which, taken hot about three or four hours after a meal, often start afresh a flagging digestion, wash the remains of the meal out of the stomach, and so prepare that organ for its needed rest.

The third rule to be remembered is: *All invalids should have their most important meal in the middle of the day.* They should only have a light meal in the evening.

All those with weak hearts have feeble digestion, because the gastric juice is both deficient in quantity and defective in quality. It is needful, therefore, in many cases, to restrict the quantity of the food, and in all to see that it is not diluted with too much fluid. Hence a fourth rule of much importance for the comfort of cardiac invalids is: *All those with weak hearts should have their meals as dry as possible.*[1]

These four rules are of great importance for all dyspeptics; but for the comfort and relief of those with weak hearts they must be strictly attended to. It may be well to recapitulate them.

Rules for feeding those with weak hearts.

1. *There must never be less than five hours between each meal.*

2. *No solid food is ever to be taken between meals.*

3. *All those with weak hearts should have their principal meal in the middle of the day.*

4. *All those with weak hearts should have their meals as dry as possible.*

A weak heart means feeble digestion; delay in digestion makes all food, and specially certain kinds of foods, prone to ferment and to break up into injurious acids and gases. Undigested food, acids, or gases in the stomach inhibit a weak heart through the pneumogastric nerve, and gives rise to intermissions, irregularity, *tremor cordis*, etc. It is of consequence, therefore, to knock out of the dietary of such patients everything likely to be difficult of digestion, such as salted, dried, or

---

[1] "In affections of the heart the most remarkable change in respect of digestion is the slowness with which liquids are absorbed by the stomach." — *A Manual of Diet in Health and Disease*, by Thomas King Chambers, M.D., Oxon., etc., London, 1875, p. 341.

otherwise preserved meats; cheese; pastry, and all similar foods in which fatty matter has undergone prolonged exposure to heat; all sweets; and nuts, which contain a quantity of oleaginous matter prone to become rancid by keeping. *Articles of food unsuitable for those with weak hearts.* Vegetable food is more apt to give rise to flatulence than animal, and all articles belonging to the cabbage tribe are specially objectionable in this respect; but such roots as carrots, turnips, and parsnips are not much better. Even potatoes require to be used sparingly. Fruits possess a low nutritive value, but when suitable they form pleasant and agreeable articles of diet when taken, as on the Continent, as a meal, such as breakfast, or as part of the mid-day meal, but they are apt to be hurtful when introduced as a mere addendum or dessert.

In treating dietetically those with weak hearts no good is to be gained by attempting to enforce rigid dietetic rules, founded upon the number of grains of carbon and of nitrogen required for carrying on the operations of life. We have to consult with our patients as to what *can*, rather than to lay down the law as to what *ought*, to be taken, keeping always the right of veto in our own hands, as in many of these cases gouty bulimia, and long perseverance in unrestrained indulgence, have depraved the appetite, and vitiated its right of selection.

Articles of diet suitable for those with weak hearts.
With due regard to idiosyncrasy, therefore, which must always be respected, we select for such cases such white fish as whiting, haddock, skate, sole, or plaice, rejecting the coarser varieties, such as cod, etc. We also recommend meat with short fibre, such as chicken, rabbit, game, mutton, or well-grown lamb, in preference to such meats as beef, whose fibres are long and tough. As few people enjoy dinner without a potato, one well-boiled, ripe, and mealy potato may be permitted, but no more. Beyond that the only perfectly safe vegetable is spinach, in which there is not a particle of flatulence; but asparagus, leeks, onions, and tomatoes may be taken in moderation if desired. Peas, beans, and other leguminous seeds tax the powers of digestion, and must be partaken of sparingly. On the other hand, such seeds are highly nutritious, and in their green state sapid, and may be used in moderation without disadvantage; the object of having a variety not being to stimulate to excess, but to be able to replace one suitable article occasionally by another, even animal food being often very advantageously replaced by fruit or vegetables.

Some people have a difficulty in taking their

food without some fluid, but this must always be restricted to the smallest possible quantity, never *more* than five ounces with any meal, and if possible less. If water be taken with the meals, it should be sipped as hot as possible; if tea, as at breakfast, it should not be stronger than one spoonful, 100 grains, to the five ounces, and infused not longer than three minutes; coffee may be made to taste, and taken either *noir* or *au lait*. Chocolate and cocoa are too much of foods for those with weak hearts, but they may occasionally be useful if taken alone, or with a bit of dry toast only; on the other hand, the infusion of cocoa nibs makes a beverage closely analogous to tea and coffee, but of a milder and less stimulating character, and therefore more suitable for many. Alcohol is not a food for the heart, and should never be prescribed, except *pro re nata ;* but so many of our patients have been lifelong imbibers of alcohol in some form or other that we can usually only restrict and not altogether prohibit. For those, then, to whom alcohol is permitted, half an ounce of whisky, brandy, or gin may be given in three or four ounces of water twice a day, along with food; or a single glass of port or sherry, or a couple of glasses of any lighter wine, such as hock or claret: each glass to measure two fluid

Fluids at mealtimes must always be restricted.

ounces; and the stronger wines are restricted because liable to give rise to acid dyspepsia if taken in larger quantity. For the same reason champagne is absolutely forbidden as a rule. But there is so much idiosyncrasy in the action of wines that each case must be arranged for separately. The only safe form of alcohol, if such a thing can be, is pure whisky and water in extreme moderation. Small quantities of alcohol are frequently prescribed as an ordinary stimulant for a weak heart, to be taken repeatedly during the day. This is a most injurious treatment, as, though the primary effect of the alcohol is stimulating, it depresses secondarily. In ordinary circumstances it is much better to direct such a patient to take two or three sips of hot water, as hot as can be swallowed, occasionally throughout the day; this will be found to have quite as good an immediate effect upon the heart as alcohol, as I have been assured by those who have tried both, while it is entirely without any secondary ill results.

Sipping hot water an excellent stimulant for a weak heart.

While desirable to keep the meals of those with weak hearts as dry as possible, it is equally needful that a sufficiency of fluid should be ingested to maintain metabolism and keep the secretions, especially those of the skin and kid-

Fluids in moderation may be safely taken between meals.

neys, in good working order. The daily allowance of fifteen ounces permitted to be taken with the food, together with the amount of water contained in the food itself, will be found to be quite sufficient to provide for all the necessary tissue changes. But if thirst be complained of, half a pint of hot water may be sipped about four hours after each meal, or only after the principal meal; this will wash all the *débris* and refuse acids out of the stomach and prepare it for its rest. Taken thus, on an empty, or nearly empty, stomach, water is readily absorbed, passes straight to the kidneys, and is not liable either to raise the blood pressure or to embarrass the heart, if taken in moderation. Hot water, as hot as can be sipped, quenches thirst much better than cold, which is of little avail. Small bits of ice to suck are also useful, but the tepid water resulting from the melting ice must be spat out, as, if swallowed, it sickens. It is often agreeable for such thirsty souls to suck a slice of lemon, and they find it useful. After all, thirst usually depends upon the catarrhal dyspepsia so commonly present in all such cases, and it ceases shortly after the dietary has been carefully regulated.

To relieve a weak heart, we must not only keep the meals dry, but it is also needful to limit the quantity of solids. For many years I have been in the habit of prescribing the following dietary, as one useful to begin with, generally sufficient, and which can be easily modified if this be found needful:—

Quantity of solid food to be allowed.

Breakfast, 8.30: One small slice of dry toast, weighing about an ounce and a half, with butter; one soft-boiled or poached egg, or half a small haddock, or its equivalent in any other fresh white fish; with from three to five ounces of tea or coffee, with cream and sugar. If there be any difficulty about the tea, it may be replaced by a similar quantity of infusion of cocoa nibs, or milk and hot water, or cream and seltzer water. Some prefer oatmeal porridge, with milk or cream, and in ordinary circumstances this need not be objected to, provided not more than four or five ounces of milk be taken, and the porridge be not more in quantity than three or four ounces of oatmeal, well boiled; provided, also, that porridge alone be taken, and not porridge first, followed by tea, toast, etc., which is destructive of all comfort, both for stomach and heart.

The *principal meal* of the day, whether it is called *lunch* or *dinner*, should be taken about 1.30 or 2 o'clock, and may consist of two courses, not more — fish and meat, or fish and pudding, or meat and pudding. *Soups, pastry, pickles, and cheese are absolutely forbidden*. White fish and meat with short fibres are preferred. Half a haddock, or its equivalent in any other white fish, boiled in milk, steamed, or broiled, never fried; wing and part of the breast of a chicken, or its equivalent in sweetbreads, tripe, rabbit, game, or mutton; one single potato, or a little spinach. For pudding, any form of simple milk pudding may be taken, or about half a pound of such fruits as pears, apples, grapes, etc., either cooked or uncooked. During this meal four or five ounces of hot water may be sipped if desired.

From 5 to 6, three or four ounces of tea may be taken if desired, infused as in the morning, not longer than four minutes, and with cream and sugar if wished; but *no* solid food must be taken with it, not even a morsel of cake or biscuit. If there be any difficulty about the tea, four or five ounces of hot water may be substituted for it, and if there seem any need for a stimulant at this time, a teaspoonful of Liebig's extract of beef may be stirred into it.

Supper, or the last meal of the day, must always be a light meal. It should be taken about 7, and may consist of white fish and a potato, or toast, with butter, or some milk pudding, or bread and milk, or Revalenta, made with milk or with Liebig's extract of beef. At bedtime, four or five ounces of hot water will soothe the stomach, promote sleep, and pave the way for a comfortable breakfast next morning.

On such a dietary a weak digestion from a feeble heart will gradually recover its tone, and the patient will feel comfortable, instead of being puffy and oppressed after meals, with an irregular and tumbling heart. The patient usually loses weight at first, from the circulation recovering its tone and reabsorbing the œdematous soakage, which, spread over every interstice of the body, often amounts to a good few pounds before it makes itself in any way perceptible as a localized œdema.

Those who have been slowly wasting from impaired digestion gain flesh from the improvement in this function, while an obese person gets thinner, from the cutting off of excesses and the diminution of the fluid taken with each meal.[1] In both an equilibrium is established as soon as the average normal weight is reached. Should this not be the case, we must first ascertain that the dietary has been strictly followed, and then proceed to alter it in the necessary

---

[1] By means of a similar dietary the patient whose pulse tracing is given at Fig. 4, p. 49, was brought down comfortably from over 20 stone to under 14 stone, in spite of his dilated heart.

direction. If the patient has been losing flesh too rapidly, the diet must be made more nutritious; this is seldom required. On the other hand, obesity may be slow in decreasing, and this will only require a little more self-denial, especially as to fluids. A mixed diet is always best for the maintenance of health, and if animal fats are not too much indulged in, the carbohydrates in the diet indicated will not be found too much.

Those who are above the normal weight, and are troubled with breathlessness, or other symptom referrible to the heart, are often set down as having fatty hearts, and no doubt they have of a kind. A heart is said to be fattily degenerated when the protoplasm of the fibres composing its myocardium becomes converted into fatty granules by retrograde metamorphosis due to defective nutrition. This arises from various causes; it occurs in connection with fevers, and other diseases, such as pericarditis, etc. It is chiefly of importance in connection with the senile heart, because it is so often found in connection with atheromatous disease of the coronary arteries, and associated with angina. It is absolutely impossible to diagnosticate fatty degeneration of the heart; we may surmise its existence, but we can only be certain of its presence when

Diagnosis of a fatty heart.

we see it *post mortem*.[1] We are often told that there is danger in treating a fatty heart, as forcible excitement of the healthy part of the fibre might tear it from its connection with the diseased portion. But a dread of this kind would hamper us sadly in the treatment of weak, dilated, aged hearts, as the signs and symptoms which those present are precisely those upon which we are told to rely in diagnosticating a fatty heart.

At page 216 will be found narrated the case of an old gentleman of seventy-seven, whose heart was diagnosed to be fatty by one of the ablest observers of his day. Yet the result of treatment was a cure, proving that a heart supposed to be fatty was only weak, and that a life supposed to be over only wanted the fillip of a few minims of digitalis to carry it on to almost the extreme of human longevity.

Cure of supposed fatty heart.

[1] "Die einfache Erfahrung, dass man in vielen Fällen von Herzdilationen mit starker Unregelmassigkeit des Pulses bei der Section oft nur eine sehr geringe oder gar keine Fettmetamorphose findet, während schwere Verfettungen der Muskelatur ohne alle Symptome von Seiten des Herzens verlaufen können, die Erfahrung also dass die Muskelverfettung nicht in directen Verhältniss zur Schwere der klinischen Symptome steht, zwingt uns ein besonderes Krankheitsbild 'Fettherz' aufzugeben." — Fraentzel, *Die idiopathische Herzvergrosserung*, Berlin, 1889, S. 191; also Balfour, *op. cit.*, pp. 309 and 348.

In true fatty degeneration no benefit can be expected from treatment, but I have never seen any detriment follow treatment, even when the heart was ultimately found to be actually fattily degenerated.

There is, however, still another form of fatty heart in which treatment may be of the greatest possible service, or the reverse, according to its character. I refer to those who are obese, whose hearts are oppressed with fat, overlying the base and infiltrating the myocardium as an *adipositas cordis*, the muscular fibres themselves remaining healthy. These hearts are usually somewhat dilated and hypertrophied, occasionally intermittent or irregular in their action. Careful dieting, cardiac tonics, rest at first, and regulated exercise subsequently, speedily improve these hearts. But obese, gouty, and breathless, without marked cardiac disturbance, these are just the cases apt to get sent off to some Spa, such as Marienbad, Kissingen, or Tarasp, to get dieted and washed out, often with the most disastrous results,[1] the treatment usual at such Spas precipitating and increasing the dilatation it is our object to avert or remedy.

Danger of Spa treatment in adipositas cordis.

---

[1] *Vide Edinburgh Medical Journal*, January, 1890, p. 607; and Fraentzel, *op. cit.*, S. 102.

When there is anasarca, or any evidence of soakage in any depending part of the body, it is of the greatest importance to place the patient, for a time at least, on the driest possible diet, and not too much of it. This is carried out by allowing for —

A perfectly dry diet of great importance when œdema is present.

Breakfast: One single slice of dry toast, weighing about an ounce and a half, with no butter, but with a single cup of tea infused not longer than four minutes, with cream and sugar, amounting in all to not more than four ounces; and nothing else.

Dinner: Not more than the lean of two chops, or its equivalent in chicken or fish; no vegetables; as much dry toast as may be desired; half an ounce of brandy, whisky, or Hollands, in three ounces of water; and nothing else.

Supper: As much dry toast may be taken as is desired, along with half an ounce of brandy, whisky, or gin, in three ounces of water; and nothing more.

It is not very desirable that a patient in this condition should drink much, even between meals, but, if thirsty, the patient may be permitted to sip slowly three or four ounces of hot water about an hour before each meal.

The relief obtained by this strict diet is both remarkable and immediate; I have seen a considerable amount of œdema of the lower limbs disappear within twenty-four hours, before there had been time for any change in the heart, which was feeble and dilated.

Most of our patients have been very self-indulgent, and are prone to assail us with loud complaints of being starved. They scarcely realize that both life and comfort depend upon strict adherence to the regulations laid down, and even while benefiting by the diet, are anxious to have the rules relaxed. "O doctor," said a lady whose feeble, irritable heart had long been a trouble to herself and me, "I have no heart now; mayn't I have a scone to afternoon tea?" "Certainly, if you wish it, but you will suffer for it." "Ah," she said, "I know that, for I have tried it." As to the starvation part of the matter, there have been so many exhibitions of fasting men of late years, that for the first week or two even the most unreasonable may be easily controlled. By that time our point will have been gained, and the improvement will be so great that we will scarcely require to appeal to the experience of Luigi Cornaro. This Venetian gentleman of the seventeenth century, after a youth of excess which destroyed his health, restored himself, after the age of forty, to perfect health by a most rigid diet. Cornaro restricted himself to a daily allowance of bread, meat, and yolk of egg, amounting to twelve ounces in all. With these solids he also took fourteen ounces of a light Italian wine each day. Upon this abstemious diet Cornaro lived in perfect health, both of body and mind, for more than sixty years, dying at last at the age of over one hundred years.[1] The only illness recorded after the adoption of this hermit fare was due to an excess of a couple of ounces in the day, both of solids and of fluids, which Cornaro was persuaded to indulge in at the instigation of friends, but to his own serious detriment.[2]

Cornaro's diet, upon which he lived in health for sixty years.

We must always, however, blend judgment with knowledge, and by occasional weighing, see that our patients do not lose weight too rapidly, and that they maintain an equilibrium when the normal has been gained. Should weight under such circumstances still continue to be lost, enquiry must be made, and, if needful, some change made in the dietary.

Tobacco is so much used nowadays that any system of dietary would be incomplete which took

---

[1] *Sure Methods of Attaining a Long and Healthful Life.* By Lewis Cornaro, London, 1820, 23d edition, p. 32.

[2] Cornaro, *loc. cit.*

no note of this. Snuffing and chewing are both so little used, in this part of the world at least, that nothing need be said of either. Smoking tobacco is so common *As to the use of tobacco.* a habit, and one so often indulged in to excess, that some rules seem requisite by which the habit may be regulated. First of all, it may be noted that the prevalent habit of cigarette-smoking and inhaling is the most seductive as well as the most injurious method of using tobacco, besides being a habit which seems most difficult to break. The only benefit ever claimed for tobacco — as a luxury — is that in some it soothes, and removes exhaustion, listlessness, and restlessness, when these are brought on by mental or bodily fatigue. But there are many who experience no such effect, and who have no excuse for the habit save imitation in the first instance and the force of habit afterwards.

Tobacco is a most potent narcotic poison; in excess it may cause sickness, vomiting, and sometimes prolonged lethargy; its action on the heart is exerted through *Tobacco most dangerous to a weak heart.* the vagus, which it first stimulates and then paralyzes. The stomach and brain are most apt to be affected by tobacco when swallowed; smoking chiefly affects the heart. At universities and schools of medicine, where young men congregate and teach each other the habit of smoking, there is always ample opportunity of studying the effect of tobacco on the heart. The tobacco heart is neither a functional nor an organic complaint; it is an acute or chronic poisoning of the vagus, which may lead to actual dilatation of the heart, and even to death itself. The tobacco heart is revealed by many vagaries, from an acute attack of intermittence, following prolonged smoking, and disappearing in a few hours, to prolonged irregular action, violent tachycardia, lasting sometimes for days; or even sharp attacks of angina, following smoking, and occasionally severe enough to prove fatal (*vide antea*, p. 127).

Narcotics in every form damage a weak heart, and are too often the cause of its debility; hence we must enquire into the habits of every patient, and endeavour to eliminate those which are

Narcotics in any form injurious to a weak heart.

This being the state of matters liable to be induced, even in young and healthy hearts, by the abuse of tobacco, it may be readily understood that elderly people with feeble hearts ought to be very chary even as to its moderate use, and on the first appearance of any sign of tobacco poisoning, such as cardiac intermission or irregularity, the habit ought to be dropped at once if any comfort in the future is desired.

injurious; not always an easy matter, as some cling to habit with an intensity which overrides even the love of life. Of no habit can this be more truly said than of the abuse of opium in all its many forms. To attempt to restore the dilated heart of an opium-eater who will not forswear his habit is even more hopeless a task than to make the same attempt in the case of a beer-soaker or an inveterate dram-drinker. I well remember telling a gentleman whose dilated heart I had for some time been treating in vain that I was sure he had not been quite open with me, as I could not obtain the expected result from the remedies prescribed, and that I was quite certain I would not have been disappointed but for the existence of some unrevealed obstacle. This appeal to his conscience produced the not unexpected confession, "Well, to tell you the truth, I take a good deal of whisky at night." The obstacle once revealed and removed, the happiest results, I am glad to say, speedily followed.

CHAPTER XI

THE THERAPEUTICS OF THE SENILE HEART. DRUGS LIKELY TO BE USEFUL, AND HOW TO USE THEM

ALL the various symptoms connected with the senile heart may be looked upon as indicating cardiac failure, with sequential complications, and the treatment must therefore be tonic, with certain modifications.

The drugs useful as cardiac tonics are but few in number, but of great value.

Digitalis is the foremost of all cardiac tonics. It gives its name to a whole group of remedies with similar actions, only one of which comes within a measurable distance of itself in the possession of valuable and reliable properties. An indigenous drug of the very highest value, and known for more than a hundred years as a most reliable remedy in dropsies, its action was so little understood, even so recently as twenty years ago,

Digitalis facile princeps of cardiac tonics.

that it was called the opium of the heart, and looked upon as a most powerful and dangerous sedative.[1] And even yet the profession are more or less hampered in its use by an idea that possesses it that digitalis is dangerously cumulative. Digitalis, like Fitz-James' blade, is both "sword and shield," and he who understands its use will never be disappointed by it, the very so-called cumulative action being but the necessary result of one of its most valuable properties when overdone. Given in full doses, at short intervals, digitalis, like many other drugs, is not wholly eliminated during the interval, but each succeeding dose reinforces those that have preceded, till a dangerous degree of cardiac contraction may be produced.[2] For this we should not blame the drug, but the prescriber. Even a considerable degree of digitalis contraction does not, however, seem to be dangerous if wittingly produced and carefully watched. It takes a good deal of digitalis to bring a human heart to a standstill in systole. Half-ounce doses of the tincture of digitalis used to be given safely and repeatedly in the treatment of *delirium tremens*. I, myself, have often successfully given drachm doses of the tincture every hour, for four or five times, in the

---

[1] *Vide Edinburgh Medical Journal*, February, 1870, p. 743.
[2] Fothergill, *On Digitalis*, London, 1871, p. 5.

precritical collapse of pneumonia; and many years ago, in treating the dilated heart of a young chlorotic girl, I kept her pulse for days at 40, and her heart-sounds beating with the empty tic-tac of an infant's heart (embryocardia). In this case all my endeavours failed to contract and cure this dilated heart, which always relaxed the moment the dose of digitalis was reduced, apparently from sheer want of tone in the muscle. Persistent treatment, though it failed to contract the heart, yet sufficed to feed it. It has kept well fed all these years, and, though a loud systolic murmur still indicates the continuance of dilatation, the patient has long been a happy wife, and the mother of several healthy children, with no appearance of any ailment about her.

What we were, perforce, reduced to in this case is all we should ever attempt in the case of senile hearts. We need never attempt to contract and cure a senile dilated heart. It cannot be done, so there is no use trying. But we can always improve the nutrition of the dilated myocardium, and in doing so we gain two ends: we fit the muscle for the more perfect discharge of its function, and we enable it better to withstand injurious influences, reflex or other.

How to use digitalis in senile hearts.

With this object in view, we employ only moderate doses of digitalis, doses which never seem to have any cumulative action, or so rarely and slightly that we may safely continue them for a week or two without observation and without risk.

These doses are for the British Pharmacopæia preparations:—

The infusion, half a measured fluid ounce.

The tincture, ten minims.

Doses of digitalis which have no cumulative action.

Each of these doses is equivalent to a little more than one grain of the powdered leaves, so that this may be taken as the medium dose that may be safely administered *every twelve hours*, without risk of cumulative action. This means that within that space of time the quantity of the drug ingested has been completely balanced by that excreted, only the tonic influence remaining; that is, the improved nutrition of the myocardium due to the action of the drug while being slowly excreted. I have known such doses to be continued for many months, sometimes for years. The dose of digitalis is not, however, an absolute one, but is relative to the bulk (weight) of the individual, and specially to the amount of his blood, a weakly anæmic individual tolerating only a very much smaller dose than one more plethoric. Now and then, too, we come across an idiosyncrasy which either tolerates freely a larger dose, or resents any but the smallest. Such cases are, however, rare; still, in view of their occasional occurrence, it is well that a patient under treatment for the first time should be seen now and then for the first week or two; afterwards, when the measure of toleration, as we may term it, has been ascertained, this may be less necessary.

There is a French preparation of digitalin, prepared by Nativelle, which is most convenient and reliable. It is made up in granules, each containing one-quarter of a milligramme (0.003858 of a grain) of crystallized digitalin. Nativelle's crystallized digitalin is said by Brunton to consist chiefly, if not entirely, of digitoxin,[1] a principle having a precisely similar action, but insoluble in water, and only sparingly so in alcohol. Be this as it may, twenty years' experience enables me to say that it is now, and always has been, a thoroughly reliable and active drug. One flaçon containing sixty granules in two months' time produces quite a decided difference in the heart-beat of those to whom they have been administered. One granule every night at bedtime is a perfectly sufficient dose to produce this decided tonic effect on the heart, and such a dose may be continued as long as may be thought necessary. Now and then a larger dose seems indicated, and one granule may

---

[1] Lauder Brunton's *Pharmacology*, London, 1891, p. 995.

be given every twelve hours, but except in bulky or plethoric individuals so large a dose as this is rarely long tolerated. To give more than two granules in the twenty-four hours is almost certain to induce speedy intolerance of the drug, and as a rule violent sickness. Occasionally even one granule in the twenty-four hours is too large a dose, and produces uncomfortable sensations. In one such case a single granule every forty-eight hours proved quite an efficient dose, and as his health improved, this patient was afterwards able to continue with one granule every twenty-four hours for several years. If preferred, Nativelle has a syrup of digitalin which contains one-quarter of a milligramme in each drachm, and by using it the dose may be subdivided as minutely as may be desired.

The object we have in view when using digitalis in a case of senile heart is not to remove dropsy, to slow the rate of pulsation, or to contract the cardiac cavities, but by the gradual accumulation of trifling advantages to tone up and strengthen the cardiac muscle by improving its nutrition. Gradually the heart acts with more vigour, the circulation improves in steadiness and force, any œdema occupying the tissue spaces is removed, and thus the blood pressure is lowered

The object we have in view in using digitalis in a case of senile heart.

and a considerable strain taken from the heart.[1] For this purpose only moderate doses are required, doses which can be continued for many months without any risk of dangerous accumulation, and which yet have a decided effect in strengthening the heart, improving the tone and elasticity of its muscle, and accumulating energy in its ganglia. Naturally this process is a slow one, and the benefit is not for a time very obvious to the recipient. Some years ago a friend called on me and said, "Doctor, your medicine is doing me no good." "Of that," I said, "you must allow me to be the best judge." "But I feel no change in my symptoms, nor any action whatever from what you have given me." "I expected nothing else; you are too impatient," I replied. "Were I to give you medicine in such a dose as to produce a sensible action in a few days, before long its action would be so unpleasant that you would either stop it yourself, or your ordinary medical attendant would order you to give it up. In a short time the seeming benefit would vanish, and you would abuse me for having given you medicine which did not agree with you, and which gave you no permanent relief. Now, what I have given you will not speedily relieve you; but give it time, and it will make you well, and prolong

[1] *Vide* Hamilton's *Pathology*, Vol. i., pp. 630 and 694; also *Edinburgh Medical Journal*, September, 1889, p. 213.

your days in comfort. Two or three months after this you will say to your wife some morning, 'Do you know, my heart is not so troublesome as it was; I begin to think I am improving'; and six or eight months after this you will come to me and say, 'Doctor, I was preaching last Sunday and feel none the worse for it.'" And so it fell out; my friend and his senile heart are nowadays, after the lapse of five years, still very good company to each other, which for many a day they were not.

The senile heart owes its peculiar symptoms and progress to the difficulty which a weakened myocardium finds in maintaining the circulation in the face of the permanent obstacle presented by rigidity of the arterial walls. To seek to excite a heart to more powerful action in the face of such an obstacle *Digitalis cannot be safely given to senile hearts without simultaneously unlocking the arterioles.* seems fraught with danger; and we know, indeed, that even moderate digitalis stimulation in such circumstances is apt to be followed by a worsening of the symptoms, sometimes by an increase of the cardiac dilatation, always of its erethism. Some also object to the use of digitalis when the arteries are atheromatous, from a dread of rupturing their brittle coats. This last-named danger must be a very infinitesimal one, as such an accident is quite unknown to me, notwithstanding a continual and free use of digitalis. But, indeed, the same means of necessity taken to prevent the increase of cardiac erethism would also prevent this more serious danger. To this end it is needful in all such cases to combine the digitalis with some drug capable of unlocking the arterioles, and of promoting the free passage of the blood to the veins. These drugs are, iodide of potassium, all the nitrites, of which nitrous ether, nitrite of sodium, and nitro-glycerine are those most commonly used. Digitalis ought never to be prescribed in a case of senile heart without the addition of one or other of these vascular stimulants, and of these iodide of potassium is the most generally useful, acting well and persistently in a moderate dose, and free from any objectionable effect.

If, at the commencement of treatment, the heart has been much neglected, the dilatation considerable, and the irregularity great, it is very desirable to begin with larger doses of digitalis than those just recommended, so as to gain control over the rate and rhythm of the heart as rapidly as possible; but these large doses are not likely to be required for any length of time, and ought to be pretermitted for at least a couple of days before the patient is put upon the smaller

Large doses of digitalis sometimes requisite, even in case of senile heart.

doses for a continuance. Where there is œdema of the lower limbs, a perfectly dry diet with tonic doses of digitalis are often quite sufficient to remove the fluid, and that in a very short time. But if the dropsy is at all considerable, it must be treated as an ordinary case of cardiac dropsy, and in such cases it is a great saving of time to drain the limbs. In all senile hearts, whatever their character or special symptom may be, we must always remember that digitalis uncombined with one or other of the vascular stimulants is never so beneficial as when it is so combined, is certain, indeed, to produce discomfort, and is very likely to do serious damage.

The only other member of the digitalis group which has any pretensions to rival digitalis itself, is strophanthus and its active principle strophanthin. Strophanthus is, however, so much more uncertain in its action, especially as to its feeding or tonic properties, than the leading member of its group, that I have never felt inclined to displace our own pre-eminent and indigenous drug in its favour. Strophanthin possesses, however, two advantages over digitalin: it is readily soluble in water, and it seem to act with great rapidity. There are, therefore, conditions in which strophanthin is to be preferred; but these are unusual and excep-

Use of strophanthus.

tional at all times, and are rarely found in connection with the senile heart.

Nux vomica is an excellent tonic for the senile heart and its concomitants, but as its usefulness depends upon its active principle, it is more advantageous and contributes to greater accuracy of dosage to employ the *liquor strychninæ hydrochloratis* rather than any of the cruder preparations. The maximum benefit is only to be got from any drug by using the maximum dose for a sufficient length of time; and to do this safely with any remedy, but especially with so powerful a drug as strychnine, it is needful to be both accurate in the dosage and regular in the times of administration. Strychnine is cumulative in its action, but by strict adherence to the rules laid down, it may be used continuously and safely for many years. I have known five minims of the *liquor strychninæ* ($\frac{1}{26}$ of a grain of strychnine) to be taken twice a day for over ten years with the very best results; at the end of that time symptoms of saturation began to appear, and the strychnine had to be discontinued. But it was no longer required; the puny, delicate, middle-aged woman is now both strong and healthy. It is only rarely that a larger dose than five minims of the *liquor strychninæ* can be given daily with benefit; three

Strychnine an admirable tonic for the senile heart.

such doses, fifteen minims instead of ten in the day, are generally followed by symptoms of poisoning in no long time. Idiosyncrasy occasionally turns up, and for this we must be prepared, but the dose indicated is the maximum dose administrable to by far the larger number of mankind, for any length of time at least. In anæmic patients there is often an intolerance of strychnine, and if employed at all, it has to be given in almost infinitesimal doses. Strychnine acts in two ways: it is an admirable tonic for the stomach, especially in those catarrhal conditions accompanied with venous congestion, so commonly present when the circulation is feeble. In this way the digestion is improved and the blood enriched, so that the body generally, and the heart in particular, gets better nourished. Strychnine has also a specially stimulating effect on the nervous system generally; consequently it stimulates and renders more excitable the vaso-motor centre, and the cardiac ganglia, probably even energizing that primordial power of spontaneous movement possessed by the cardiac muscular fibre itself — a power which may be looked upon as a remnant of the *vis insita*, the once diffuse nervous force. In virtue of this action on the heart and nerve centres strychnine increases the cardiac force, raises the intra-arterial blood pressure, and is —

next to digitalis — the most excellent tonic we possess for all feeble and dilated hearts. In the less serious class of cases it is sufficient of itself to give tone both to the heart and the system generally, while in the most serious cases it is a most useful adjunct to digitalis.

Arsenic is another of our most valuable tonics. It is advantageously employed in many forms of disease, and it is quite indispensable in the treatment of the senile heart.

Uses of arsenic.

It is very useful in those congestive conditions of stomach which accompany cardiac failure; and its effect in angina is sometimes almost magical, the suffering disappearing like a dream, quite apart from any influence exerted on the cardiac failure upon which that suffering seemed to depend. Masselot and Trousseau have both remarked upon the increased capacity for exercise that follows the administration of arsenic,[1] and this doubtless depends upon the same general tonic influence, affecting the lungs, heart, and blood, that makes breathlessness a thing unknown to the Styrian mountaineer, and restores the blooming coat and friskiness of youth to old

---

[1] "J'insiste sur ce phénomène eprouvé également par M. Masselot, et signalé par lui en ces termes : 'très grande aptitude à la marche.'" — *Traite de Therapeutique*, par A. Trousseau et H. Pidoux, Vol. i., p. 312.

and seemingly worn-out horses. The Styrians are accustomed to take large doses, as much sometimes as five grains, of pure arsenious acid[1] in the day, but such dangerous doses are by no means necessary to obtain the tonic benefits of the drug. Most excellent results indeed occasionally follow the prolonged use of almost infinitesimal doses. I well remember one old gentleman, exceedingly sensitive to the action of drugs, to whom the $\frac{1}{50}$ of a grain of arsenious acid was quite poisonous, but who could tolerate the $\frac{1}{100}$ of a grain without difficulty. After taking this minute dose daily for two or three weeks, and nothing else, for a dilated and hypertrophied heart beginning to fail, he said to me, "I don't know what benefit you expected from the treatment, but I know what I have received; I can go upstairs much easier than I used to do." Arsenic may be given alone, and in anæmic and very sensitive patients who can only tolerate a very minute dose this is often the best way of employing it. To these one granule of arsenious acid containing $\frac{1}{50}$ or $\frac{1}{100}$ of a grain may be given after food once or twice a day for many months with only increasing benefit. More usually

[1] Dr. R. Craig Maclagan, "On the Arsenic-eaters of Styria," *Edinburgh Medical Journal*, September, 1864, pp. 203, 206.

it is better to combine the arsenic with digitalis or strychnine, or with both. In making any of these combinations, the *liquor arsenici hydrochlorici* is the better preparation to employ, and in combination, with the *liquor strychninæ hydrochloratis* it is the only one that ought to be used, as with it the other preparations, the *liquor arsenicalis*, or the *liquor sodii arseniatis*, make an incompatible and more or less unsightly mixture. As we learn from the histories of the Styrian arsenic-eaters, arsenic is a poison to which the system may be gradually habituated, so that even large doses may be taken for many years, not only with impunity, but with positive benefit. When given therefore in the moderate medicinal dose of two or three minims of one or other of the fluid preparations, we may safely continue them twice a day for as long as we think needful, without any misgivings. Nor need we have any dread of any danger in leaving off the drug after long continuance, as was at one time alleged. A little caution may be required in commencing its use, as idiosyncrasy plays a marked part in relation to arsenic, but it is only rarely that we fall in with those who are extra-sensitive to its action.

When the blood is deficient in hæmoglobin, iron is a necessity. It is best given along with food, and should never be combined with digitalis, as such a combination is very apt to sicken. The proto-salts of iron are to be preferred to the per-salts, as they are more easily decomposed by the acids of the gastric juice, and are thus more readily assimilated. As a rule, large doses are not required in cases of feeble or dilated heart. *Iron useful when hæmoglobin deficient.*

Intra-arterial blood pressure depends upon the distension of the arterial system by the blood contained within it. This vascular turgescence, in its turn, depends upon the relation between the amount of blood pumped into the arteries by the heart and the outflow through the arterioles. After middle life the outflow through the arterioles is hindered by obsolescence of the capillaries and by loss of arterial elasticity, and the blood pressure is raised by these obstacles, even with a heart beating at its normal rate and force. A healthy heart has sufficient reserve force to enable it to cope successfully with the demand for extra exertion thus made upon its powers, and it thrives upon its exertion. But when the heart is from any cause feeble or ill-fed, it fails to respond, and it suffers, its suffering giving rise to those varied symptoms comprised under the term "senile heart" (*vide antea*, pp. 27, 35, etc.). *What blood pressure actually is.* *A high blood pressure originates trouble to the senile heart.*

T

As these sufferings are caused and maintained by the high blood pressure, whatever lowers this always gives relief. Hence these sufferings are capable of being relieved by various modes of treatment which are not all of them truly remedial.

Whatever lowers the blood pressure gives relief.

Permanently to remove these sufferings, we must not content ourselves with merely reducing the blood pressure; we must also so strengthen the heart as to enable it to cope with a blood pressure always over the normal of adolescence, and which is liable to be suddenly abnormally raised by many causes. Cardiac tonics are, therefore, required. But all cardiac tonics — except, perhaps, arsenic — are also cardiac stimulants: they increase the elasticity and contractility of the heart, and, in certain conditions, they improve the heart's metabolism by enabling it to feed itself with a larger blood-wave at a higher pressure. When the heart is feeble, however, this is just what cannot be done. The whole trouble has arisen because the blood pressure is already too great for the powers of the heart, and if we goad this feeble organ to further exertion, for which it is unfit, we either increase any dilatation that may be present, or induce erithistic tachycardia or irregularity. Cardiac tonics don't agree; but we can make them agree by combining a vascular stimulant with a cardiac stimulant; then things work smoothly. The heart, no longer opposed by an obstacle it can either not overcome, or only imperfectly and with suffering, now contracts more perfectly, feeds itself better, and all its sufferings vanish.

Importance of the combination of vascular stimulants with cardiac stimulants in the treatment of the senile heart.

Vascular stimulants are agents which dilate the peripheral vessels (arterioles) and so promote the flow of blood from the arteries into the veins and lower the intra-arterial blood pressure. Iodide of potassium is not, perhaps, generally regarded as a vascular stimulant, but in so far as it promotes the flow through the arterioles, and lowers the blood pressure, it is an eminent member of that group, as has been established experimentally, and duly recognized in relation to the treatment of aneurism.[1] It is not rapid in its action, but it is persistent, two or three grains every twelve hours being quite sufficient to enable digitalis to be given freely without any cardiac disturbance.

All the nitrites are vascular stimulants. Spirit of nitrous ether or nitrite of sodium may be so

---

[1] *Vide* Bogolepoff, Zur Frage der Physiologischen Wirkung des Iodkalium, *Moskauer Pharmacolog. Arbeiten*, S. 125; and Virchow's *Jahresbericht*, 1876, erster Band, S. 402; also Balfour, *op. cit.*, second edition, p. 459.

employed, but their action does not last so long as that of the iodide of potassium; while in rapidity they are far inferior to either the nitrite of amyl or nitro-glycerine. Besides being a vascular stimulant, with all the actions belonging to such remedies, nitrite of amyl is also an analgesic. A former patient who suffered from intense anginous pain, accompanying a large aortic aneurism from which he died, always found the analgesic action absent unless the drug was freshly prepared. When not quite fresh, his face flushed, and all the usual symptoms due to amyl were produced, but the pain was not relieved. Since then the amyl has been retailed in hermetically sealed glass capsules, apparently with the effect of retaining the analgesic properties. The flushing of the face, the fulness of the head, and the rapid action of the heart produced by the amyl are very disagreeable to some patients; they do not seem to be in any way injurious. I have known amyl to be used with great freedom in angina. One medical friend, who suffered much from angina, connected with aortic regurgitation, not content with inhaling it frequently during the day, used to soak his pocket handkerchief in the amyl and go to sleep with it lying on his face, without any ill results. The action of the amyl is very evanescent, the smell is disagreeable, and the quantity in a single capsule is rather small, but that is a matter easily remedied.

Nitro-glycerine, glonoin, or trinitrin, is said to be a nitrate of glyceril, but its action is that of a nitrite. In ordinary medicinal doses of $\frac{1}{50}$ to $\frac{1}{100}$ of a grain it rapidly lowers the blood pressure and relieves the pain of angina. The action of nitroglycerine is somewhat prolonged, from one to three or four hours, according to the dose. By giving an anginous patient three or four doses of nitro-glycerine in the day he can often be kept quite free from his attacks; it is well to give him a dose half an hour before any exertion likely to bring on an attack, and also just before going to bed. As a one per cent solution it acts very rapidly, and the dose is from one-half up to ten or more minims. In the form of tabellæ, made with chocolate, each containing the $\frac{1}{100}$ of a grain, it acts nearly as quickly, — in about half a minute, — if the lozenge be chewed as rapidly and perfectly as possible. The drawbacks to the use of nitroglycerine are its liability to produce headache, giddiness, throbbing of the cerebral arteries, and palpitation of the heart, but it is remarkable how seldom these are complained of. In the form of tabellæ, — tablets or lozenges, — the nitro-glycerine is easily carried about, and is readily available on the slightest indication of pain.

Throughout the literature of cardiac disease there are recorded many cases of extreme and distressing irregularity of the heart at once relieved by a fit of gout; all of these would no doubt have been cured just as well and as speedily by the use of colchicum. This is a matter not to be lost sight of. The senile heart is the gouty heart, and anti-arthritic medication is always useful, sometimes of paramount importance, and it may always be combined with other necessary remedies, notably with digitalis.

Colchicum often of great service in irregularity of the heart.

In all cases of gouty heart it is of consequence to keep the *primæ viæ* free from acidity, and this of itself is often a cure for many of the cardiac symptoms, especially irregularities of rate and rhythm. I have known many who have taken with great benefit a teaspoonful of carbonate of soda, or of the bicarbonate of potash, at bedtime every night for many years. This medication has seemed to me to favour the formation of Heberden's knobs, but it has undoubtedly been attended with relief to the cardiac symptoms.

A thorough alkalizing of the *primæ viæ* is also readily carried out by the use of Vichy water, either plain or aërated. This may either be taken as ordinary drink, or a small tumblerful may be taken before break-

Importance of alkalizing the primæ viæ.

fast, while dressing, and another about an hour before dinner. A third tumblerful, on going to bed, is an excellent thing for those who are gouty and have no marked cardiac symptoms; but it is not wise for a heart patient to go to bed with a full stomach, though its contents be only water. The thorough alkalizing of the *primæ viæ* is an excellent adjuvant, and often of itself suffices to remove slight irregularity of the heart.

Dr. Gregory's gouty powder, or its modern analogue rhubarb, with the bicarbonate of potash or soda, is a most excellent antacid aperient, as I suppose every gouty person knows full well. When there are symptoms of gastric irritation, the addition of bismuth to this powder is of great advantage, and has often been successfully employed in cardiac intermittence or irregularity when accompanied by such symptoms.

An active cholagogue purge is one of the most efficient means of lowering the blood pressure and relieving the heart. This it does in virtue of the large quantity of fluid it drains from the blood, as well as by the increased amount of blood attracted to the intestinal mucous membrane by the irritation of the purgative. Any cathartic would suffice for this, but a cholagogue, or one which acts by increasing the secretion of the liver,

Cholagogue cathartics useful in lowering the blood pressure.

has the additional advantage of directly relieving the right side of the heart.[1]

Flatulence is a symptom that often produces a great deal of distress, not from mere distension, — though that, too, occasionally disturbs, — but by its action on the heart, causing intermission, irregularity, or severe attacks of *tremor cordis*. In a stomach congested and catarrhal from feeble circulation, an amount of flatulence insufficient to produce any feeling of distension often gives rise to great cardiac uneasiness and disturbance that passes off at once on eructation. These symptoms are generally of reflex origin, but in long, narrow chests flatulent distension often seems to produce cardiac disturbance by direct pressure. At least, it is not uncommon for such a patient — to all appearance, and to his own feeling, in perfect health — on stooping to pick up a pencil, or tie his shoe, to have his heart run off in a fit of irregularity or of *tremor*. So sudden and unexpected a seizure is very alarming to most. It is the one occasion upon which a sip of spirits — whisky or brandy — seems permissible. At the same time half a drachm of *spiritus ammoniæ aromaticus*, with an equal quantity of *spiritus lavandulæ com-*

Flatulence, its results, and its treatment.

[1] Lauder Brunton, *Disorders of Digestion*, Macmillan & Co., London, 1886, p. 208.

positus, in a little water will give relief as certainly and as quickly, but it is not so easily carried about as a small flask of spirits; moreover, the spirit acts best undiluted, which is handy. The treatment of flatulence demands careful dieting, apart from the special needs of the case generally, along with the persistent use of the old-fashioned cobbler's pill, the compound galbanum pill of the last Edinburgh Pharmacopeia, the *pilula assafœtidæ composita* of the British Pharmacopeia.

Narcotics are of use in the treatment of affections of the senile heart to relieve pain and to procure sleep. In relieving pain we generally also induce sleep, but there are many hypnotics well fitted to induce sleep which are of little use to relieve pain. There is only one hypnotic which is also a sure analgesic, and that is *opium*. Its alkaloid, morphia, acts so rapidly and certainly, and is so readily administered hypodermically, that it deserves every confidence. *Morphia* is a useful and reliable remedy, not only in painful angina, but also in those vaso-motor anginas which are attended by great breathlessness without pain, inasmuch as it is not merely an analgesic, but also an antispasmodic, and it lowers the blood pressure by relaxing the arterioles and so favouring the trans-

Narcotics and their use to relieve pain.

Action of morphia.

ference of the blood from the arteries to the veins. Morphia is not only an analgesic and anti-spasmodic, but also a hypnotic of great power, and as it has no ill effect, either on the heart or respiratory centre, it may, when required, be freely used for both purposes. The drawbacks to its use are the headache and gastric disturbance it is liable to produce, and also the risk of inducing the morphia habit. There is nothing, however, that can replace it in certain cases, and in them there seems but little risk of provoking the habit.

Chloroform is sometimes useful to relieve pain, when severe, till morphia has had time to act. Chloroform is analgesic and hypnotic only because it is anæsthetic. It relieves pain and induces sleep only by producing entire loss of sensibility to all external impressions, — a condition not wholly devoid of danger, and requiring to be carefully watched, as the border line of safety is so easily crossed. With careful dosage there is no risk whatever in giving it to diseased, feeble, possibly, or actually, fatty hearts.[1] The risk is not in the drug itself, but in its administration.

Action of chloroform.

Chloral, like chloroform, is an anæsthetic, and in virtue of this property it both relieves pain and induces sleep.

Action of chloral.

[1] *Vide* Balfour, *op. cit.*, second edition, 1882, p. 308.

It does not, however, act so rapidly as either morphia or chloroform, and is not, therefore, likely to take the place of either. Properly administered, it is a perfectly safe and certain soporific, and as such it has its use in certain cases. Liebreich's chloral is the only preparation always safe, and, therefore, always to be used. Fifty grains of this will put all well people to sleep, forty grains will put to sleep a great many who are not well. It is given off at the rate of ten grains an hour, so that after the lapse of forty minutes the organism still retains over thirty grains. To this, if need be, we add a second dose of forty grains. This will put to sleep a very large proportion of those still sleepless; and I have never known any one resist a third dose of forty grains, nor, I may add, have I ever seen anything but the best results from even the full dose of 120 grains. This dose, though a large one, is quite within the limits of safety, even if swallowed all at once. But, given in the manner prescribed, there is in the organism at the end of 120 minutes little over 100 grains. Given in this way, chloral is a perfectly safe and perfectly certain hypnotic, and there are cases even of heart trouble in which this knowledge may be useful.[1] Both chloral and

---

[1] Richardson says that a man weighing 120 to 140 pounds is thrown into a deep sleep by ninety grains of chloral, and that

chloroform lower the blood pressure by causing dilatation of the arterioles, probably by paralyzing the vaso-motor centre.

There are three hypnotics, pure and simple, which deserve attention in those many cases of insomnia which so often accompanies a gouty and feeble heart. The first of these is *paraldehyde*, a very reliable hypnotic without any analgesic properties. Under its use the blood pressure falls from paralysis of the vaso-motor centre, but the heart seems to be unaffected. Paraldehyde may be given in considerable doses, as much as a drachm every hour, till sleep ensues. The great drawbacks to its use are, its vile taste, which may be overcome by giving it in an aromatic mixture or in a capsule, and the disagreeable odour which the patient exhales for the twenty-four hours following its ingestion, which nothing seems able to remedy.

Action of paraldehyde.

The second hypnotic deserving of attention is *chloralamid*. This is an excellent soporific: it lowers the blood pressure, but it also quickens the heart-beat, and is thus inferior in usefulness to the third hypnotic to be spoken of immediately. In spite of this drawback, chloralamid may be

Action of chloralamid.

the sleep which follows 140 grains is dangerous. — *Journal of Mental Science*, Vol. xviii., p. 118.

given quite safely to cardiac patients for many weeks. It is not cumulative in its action, nor does use ever seem to necessitate an increase of the dose. Forty grains is an efficient dose. This should be rubbed up with spirit (0.920 sp. gr.) *donec solutio fiat*, and taken with the addition of a little syrup.

The third hypnotic of importance is *chloralose*, a drug of quite recent introduction, but which promises to be a most valuable addition to our armamentarium. Chloralose lowers the blood pressure; but, even in large doses, it has no exciting effect on the heart, seeming rather to steady and regulate the action of that organ. Patients fall asleep quickly under its use, and they waken easily and refreshed. There is no headache and no gastric disturbance, the appetite seeming to be rather improved. Even when taken in an excessive dose, the heart is never affected injuriously, the only result of an overdose being a certain amount of intoxication. The only drawback to the use of chloralose is a tendency to act irregularly, and to induce nervous symptoms in hysterical and neuropathic patients. The dose of chloralose is from two to eight grains, and it is best administered in a cachet. A very good way of giving chloralose is to give a cachet containing three or four grains at bedtime. Should

Action of chloralose.

the patient have a good night, well and good; but should he wake after an hour or two of sleep, the repetition of a similar dose will secure a good night's rest.

The bromides are often of the greatest service, especially in the senile hearts of females about their climacteric. But the bromides are pure sedatives, and are not to be trusted to for any hypnotic action. The bromide of potassium is supposed to enfeeble the heart's action; a similar objection is not applicable either to the bromide of ammonium or of sodium.

Use of the bromides as sedatives.

CHAPTER XII

THE PROGNOSIS OF SPECIAL SYMPTOMS. RECAPITULATION OF TREATMENT, WITH SPECIAL REFERENCE TO SYMPTOMS

THERE is, strictly speaking, only one possible prognostic dictum applicable to all senile hearts; fortunately a lapse of many years often intervenes between such a prediction and its fulfilment.

When prognostication is required in reference to any special symptom, and its relation to the prolongation of life, the answer is neither so simple nor so certain.

Precordial anxiety often distresses a patient greatly. It is the earliest symptom of the senile heart (*antea*, p. 35), and the prognosis is favourable provided the cause is remediable. *Prognosis in relation to symptoms. Precordial anxiety.*

If the cause of myocardiac weakness is irremediable, or if a remediable cause is neglected, and allowed to continue its evil influence, the weakened myo- *Intermission and irregularity.*

cardium speedily comes under the influence of reflex inhibition, and the heart's action becomes intermittent or irregular. Now, a man with an intermittent or irregular heart *may* live for many years; but his life is handicapped by his heart, and if the cause of the myocardiac debility is irremediable, or is carelessly allowed to continue its injurious influence, in no long time the heart dilates (*antea*, p. 40), and the declension becomes more rapid. At any age an intermittent or irregular heart is amenable to treatment, and may be cured. But a heart dilated after middle life is, to say the least of it, only rarely rehabilitated; it has taken a downward step which is seldom retraced. Life is now more seriously handicapped; breathlessness and œdema are not long in following.

Any violent shock may force even a strong heart to intermit or become irregular. But in such a heart intermissions die away in from six months to a year (*antea*, p. 43). Any sudden shock acting on a feeble heart may prove immediately fatal, or a less severe shock, worry, or anxiety may bring on intermission and irregularity, or may precipitate serious dilatation of the heart, terminating fatally in a few months, anticipating by more than a dozen of years the natural progress of the affection (*antea*, p. 44).

Palpitation affects the young rather than the old, and though a distressing symptom, it is rarely attended by any danger (*antea*, p. 63).

Palpitation.

Tremor cordis is a most alarming symptom to the sufferer. It does occur in early adolescence, but rarely; after middle life it is common enough. It does not seem to have any marked injurious influence, and though, perhaps, not specially favourable to longevity, any effect it may have in shortening life, or even in promoting cardiac dilatation, has not as yet been ascertained. *Tremor cordis* seems to be always connected with gastric disturbance, and is rarely unaccompanied by some of the other phenomena of the senile heart (*antea*, p. 64).

Tremor cordis.

Tachycardia is always a symptom (*antea*, p. 71), and its prognosis depends upon its cause. When tachycardia has been brought on by vagus poisoning, as by alcohol, tobacco, etc., the prognosis is not serious, though there is considerable temporary risk to an aged heart. Reflex tachycardia (*antea*, p. 82) is in most cases readily curable, though it sometimes lasts for years, apparently without any detriment to the sufferer. When associated with inflammatory affections of the myo- or endocardium, the prognosis must be very guarded. It becomes less

Tachycardia.

serious when the affection gets localized as a valvulitis. The prognosis of tachycardia is most serious when it is associated with compression of vagus by a tumour.

Bradycardia. Of this there are two forms: one, the gouty variety, depends upon alternating hemi-systoles (*antea*, p. 92); and the other, the true bradycardia (*antea*, p. 106). Both varieties are associated with dilatation of the heart, but the hemi-systolic form is amenable to treatment, and its prognosis is that of an ordinary dilated heart, dependent on the age of the patient and the condition of his myocardium. True bradycardia — and the two varieties can always be differentiated by their sphygmograms — is a very serious affection, and life seems rarely to be prolonged beyond three or four years, the end being precipitated by an epileptic attack. Hemi-systolic bradycardiacs are also exposed to a similar risk, but in them this risk is never so imminent, and it may be averted.

Delirium cordis is always a serious affection. If it be impressed on a strong heart by a combination of work and worry, it may, with care, continue to handicap the sufferer for as long as twenty years. As a rule it is most likely to be found in connection with feeble, dilated hearts, and then a fourth part of that period will probably cover the termination.

Angina pectoris affords an instance in which experience enables us to give a more hopeful prognosis than professional opinion would at first be inclined to homologate. *Angina pectoris.* Every case of so-called pseudo-angina must be considered on its own merits. Hysterical angina is of little importance. In gouty angina, if the attacks are hysterical in character, it must come under that category; if otherwise, it must be considered as an ordinary angina. In every case of angina the greater the suffering of the patient, and the less there is discoverable wrong with the heart, the greater the danger, and at the most a few months will include the termination of the case. If the heart be simply dilated, treatment may be of much service, and life may be prolonged for a dozen of years. If the heart is already considerably hypertrophied before the angina sets in, treatment is never of so much service, and life is not likely to be so prolonged.

Affections of the heart, and especially senile affections of the heart, are not adapted for accurate prognosis. In all of them the element of uncertainty bulks too largely: we must therefore carefully refrain from any too dogmatic assertion. Still, it is of consequence to know the exact nature and the probable result of any special symptom, such as *tremor cordis;* and though somewhat wanting in definiteness, the foregoing statements may yet be useful to many.

Tabular Recapitulation of Treatment.

In every case careful removal of the *lædentia*.

Precordial anxiety. Careful dieting; cardiac tonics; rest at first, afterwards regulated exercise.

Intermission and irregularity. Careful dieting; vascular stimulants, combined with cardiac tonics; sedatives, especially for women about their climacteric, occasionally hypnotics; antacids and anti-arthritics; assafœtida (*pil. galbani co.*); moderate exercise.

Palpitation. Antacids; stimulants; mustard over precordial region; hot foot-baths. In interval strengthen patient by open-air exercise, good food, and such tonics as may seem needful, especially iron.

Tremor cordis. Careful dieting most important; antacids; anti-arthritics; *pil. galbani co.*

Tachycardia. Careful dieting; in recent cases following cardiac overstrain, belladonna, or atropine, must be pushed till pupils dilate. In cases of poisoning by tobacco or alcohol, tonic doses of digitalis useful. Cardiac tonics, especially digitalis and arsenic, continued for a long time in moderate doses, supplemented by hypnotics at bedtime, especially morphia. Digitalis most useful in vagus paralysis, morphia in affections of the sympathetic. Cholate of soda slows the pulse, but it destroys the blood corpuscles, and the benefit is thus a doubtful one. Antipyrine has been recommended theoretically. Faradization of the skin over the *precordia*, or of the vagus nerve; or the skin or vagus may be galvanized. Compression of the vagus. Forced inspiration, holding the breath as long as possible. Ether sprayed along the cervical spine. A chloroform poultice over the precordial region.

In the hemi-systolic variety, cardiac tonics, especially digitalis. In true bradycardia digitalis is also indispensable, to maintain the elastic tonicity of the heart, and to enable the heart to cope with the exceptionally high blood pressure (*antea*, p. 106) prevalent during part of the systole. *Bradycardia.*

Careful dieting, vascular stimulants, cardiac tonics, antacids, and antiarthritics. *Delirium cordis.*

During the paroxysm, nitro-glycerine, nitrite of amyl, chloroform and morphia. During the interval most careful and abstemious diet, especially towards evening. Vascular stimulants in combination with *Angina pectoris.*

cardiac tonics, especially arsenic. Exercise is to be avoided, and only undertaken when duly prepared for by the ingestion of some vascular stimulant.

Such, then, is the armamentarium most useful in senile heart troubles. Its constituents are all valuable remedies, and though some of them are interchangeable, yet each has its own peculiar mission for which it is best adapted. Each case must be carefully considered from every point of view, thoroughly individualized, and the treatment best adapted to attain the end in view firmly laid down and persistently carried out. A disease that has been gradually coming on for thirty or forty years cannot be expected to yield to a week or two of treatment, however skilfully devised or carefully carried out. It often takes many months of care before an irregular heart is made regular, or the declension of a failing heart is arrested. In time, however, all this can be done. Time, however, is required; for it is not to be done by any dexterous legerdemain, but by the skilful imitation of natural processes, and by the steady accumulation of trifling advantages; and our drugs must be mixed like Opie's colours — with brains.

INDEX

Aconite, action of, 81.
Action, idio-ventricular, 39.
   irregular, of heart, how produced, 40, 48.
   irregular, of heart, danger of, 40.
Adipositas cordis, 251.
Age, the result of tissue change, not of years, 19.
   alteration in arterial system due to, 13.
   typical death from, 18.
Alcohol unsafe for aged hearts, 243.
Amyl, nitrite of, 276.
   nitrite of, an analgesic, 276.
Anabolic nerve of heart, 39.
Anæmia, a source of heart trouble, 29.
Anæsthetics, 282.
Analgesic, the only real, 281.
Angina pectoris, 115.
   may occur in early life, 116.
   syndrome of, 121.
   prognosis in, 141, 291.
   cause of, 125.
Angina, vaso-motoria, 131.
Antacids, 278.
Anti-arthritics, 278.
Anxiety, precordial, 35.
Aortic second, accentuation of, 35.
   accentuation of, what it indicates, 57.

Aortic second, a booming, 57.
   regurgitation, how produced, 57.
   systolic murmur, 58.
   systolic murmur precedes regurgitation, 58.
Arrest of heart's action, voluntary, 67.
Arsenic as a cardiac tonic, 270.
   use of, by Styrians, 270.
   use of, may be continued for years, 272.
Arterial system first to fail, 19.
Arteries of the young may be rigid, 229.
Arterio-capillary fibrosis, 200.
Arthritis, rheumatoid, 181.
Asthenia, ingravescent, 29.
Atherosis, 205.
Asthma, cardiac, 135.
   death from, 137.
Asystole may be sudden or ingravescent, 80.
   case of ingravescent, 139.
Augmentor and accelerator nerves, 37.
Auricular murmur, its position and cause, 55.
   why not always to be heard, 55.
Auscultation a means of detecting cardiac dilatation, 53.

Balfour, W., his treatment of gout by massage, 172.

Bile, the mere drainage of a manufactory, 187.
   amount of, in man, 188.
   free secretion of, relieves the heart, 189.
Bismuth, 279.
Blood pressure, what it is, 273.
   in youth, 11.
   rises when growth ends, 12.
   changes in, from age, 14.
   an increase of, embarrasses the heart's action, 273.
   increased, tends to dilate the heart, 60.
   a healthy heart successfully copes with, 25.
   lowering the, relieves a weak heart, 274.
   effects of high, intra-pulmonary, 159.
   effects of low, intra-pulmonary, 160.
   effects of vascular environment on, 27.
   lowered by vascular stimulants, 275.
   lowered by cholagogue cathartics, 279.
   indications, diagnostic from high, 226.
   indications, diagnostic from low, 225.
   effects of, on arterial tension, 228.
Bradycardia, 51, 93.
   hemi-systolic, 92.
   hemi-systolic, case of, 92.
   prognosis of, 290.
   true, 107.
   Holberton's case of, 99.
   case of, 109.
Bromides as sedatives, 286.
Bulimia, gouty, 186.

Capillaries, phenomena due to obsolescence of, 14.

Cardiac movements primordial in character, 36.
   influence of nervous system on, 37.
   irritability, 220.
Case of dilatation of heart, 42.
   gouty glycosuria, 191.
   supposed fatty heart, 216.
   Colonel Townsend, 67.
   illustrative of angina, 145.
   angina in young woman, 122.
   ingravescent asystole, 139.
   irregular heart, 46.
   tachycardia, 78.
   bradycardia, 92, 93, 99, 109.
Cathartics cholagogue, lower blood pressure, 279.
Cervical cord, injury to, produces bradycardia, 98.
Chloral, hydrate of, 282.
Chloralamid, 284.
Chloralose, 285.
Chloroform, 282.
Circulation, condition of, up to middle life, 11.
Colchicum, use of, in senile heart, 278.
Cornaro, Luigi, 253.
   his diet, 254.
Cullen's definition of gout, 163.

Death, rarely due to age alone, 1.
   defined, 9.
   sudden, from emotion, 31.
   from angina, 138, 139.
   from age, typical, 18.
Decay, premature, 5.
   final stage of development, 4.
Delirium cordis, 113.
   prognosis in, 290.
Deposits of urates in gouty joints, 164.
Depressor nerve of heart, 38.
Development ends only with death, 4.
   may be precocious, 4.

Developmental phenomena may be terminal as well as initial, 4.
Diathesis, gouty, 161.
Diet, dry, 252.
    Cornaro's, 254.
Dietaries, 237.
Dietetic regulations, 238.
Digitalis, use of, 260.
    accumulation of, 259.
    accumulation, how to avoid, 261.
    object and mode of using, 263.
    must be combined with vascular stimulants, 266.
Dilatation of heart, time required to produce, 44.
    effect of, in displacing apex-beat, 53.

Emotion, intensity of, an important factor, 45.
    may prove suddenly fatal, 31.
    fruitful source of heart trouble, 31.
Epilepsy, character of attack in bradycardia, 104.
Excess in food more dangerous than in drink, 236.
Exercise, 232.
Exertion, effect of, on an anæmic pulse, 47.
    danger of unduly prolonged, to the heart, 30.

Fakeers, Indian, how they slow the heart, 68.
Fasting men, 253.
Flatulence, disturbs the heart directly, 280.
    disturbs the heart reflexly, 280.
Force, vital, what it is, 9.
    cause of failure of genesis, 10.
Fothergill's case of voluntary slowing of heart, 67.

Fermentation test not devoid of fallacy, 190.
Fluids must be restricted at meal times, 243.
    less injurious between meals, 244.

Giants, what they are, 12 (note).
Graves' disease, syndrome of, 71.
Glycosuria, gouty, 191.
    cause of, 195.
Glycuronic acid decomposes copper in Fehling's test, 190.
Gout, Cullen's definition of, 163.
    in no respect inflammatory, 164.
    resolution of paroxysm always incomplete, 164.
    temperature of affected joint, 164.
    Balfour, W., his treatment of, 172.
    Sir W. Temple's treatment of, 170.
    The Rhyngrave's treatment of, 171.
    massage in the treatment of, 170.
Gouty diathesis, what it is, 161.
    present in every one, 162.
Gouty paroxysm a mere episode in its history, 163.
    history of a, 165.
    due to infarction, 166.
Growth, influence of heredity in causing cessation of, 13 (note).
    precocious, not identical with premature development, 5 (note).
    conditions of, in early life, 11.
Hæmogenesis, interference with, 224.
Hæmolysis, causes of, 226.
Heart always hypertrophied in the old, 22.

Heart, changes in, from age, 22.
  sources of vigour in the senile, 25.
  cause of trouble in the senile, 27.
  changes in, when dilated, 52.
  idiopathic enlargement of, 89.
  gouty heart, 34.
  nervous, 34.
  innervation of, 37.
  inhibition of, 43.
  inhibition of, favours dilatation, 43.
  proportion of senile, to ordinary heart affections, 143.
  proportion of anginas among senile hearts, 144.
  proportion of anginas among males and females, 144.
  proportion of anginas cured, 144.
  essential lesion of the senile, 33.
  symptoms of the senile, 35.
  troubles of the, always alarming, 214.
  troubles rarely arise from failure of the trophic nerves, 215.
  troubles may be remedied at any age, 216.
  fatty, diagnosis of, 249.
  supposed fatty, generally only weak, 250.
  irritable, 217.
  affections often last long, 219.
Hemi-systolic bradycardia, 92.
Hyalin fibroid disease, 200.
Hyperdicrotism in tachycardia indicates danger, 79.

Infarction, what it is, 166.
  cause of gouty paroxysm, 166.
Inhibition of heart, 43.
  favours dilatation, 43.

Interference, vagus, the cause of irregularity, 48.
Irregular cardiac action, case of, 46.
  causes of, 48.
  diminishes efficacy of heart beat, 41.
  danger of, 40.
  is never unimportant, 41.
  prognosis in, 287.
Ischæmia, cardiac, its relation to angina, 126.
  causes of, 126.
  cause of pain in angina, 131.

Jenner, Edward, first to point out that ischæmia was the cause of pain in angina, 130.

Katabolic nerve, the, of the heart, 38.
Kidney, relations of, to heart, 196.
  Bright's idea of, 197.
  Traube's idea of, 197.
  George Johnson's idea of, 198.
  Gull and Sutton's idea of, 200.
  the red, contracting, 196.
  the senile, 202.
  the senile, a true gouty, 202.
  the gouty, preventable, 202.
Knobs, Heberden's, 178.
Kreatin, and kreatinic acid, decompose copper in Fehling's solution, 190.

Life defined, 8.
  form of energy, 7.
"Luxus" heart, the, 34.
  not due to overfeeding alone, 227.

Massage, treatment of gouty paroxysm by, 170.
Metabolism, danger of imperfect, 223.

INDEX

Morphia, uses of, 281.
Muscles, twittering of the, 183.
Myocardium, weakness of the, its symptoms, 33.
    failure of the, 220.
    failure of the, treatment of, 231.

Nails, ridged, 177.
    furrowed, 177.
Narcotics, danger of, to the senile heart, 257.
    use of, 281.
Nitrite of sodium, 275.
Nitrites, action of, 275.
Nitro-glycerine, 277.
Nodosities, Haygarth's, 180.
Nux vomica as a heart tonic, 268.

Obesity, how to reduce, 248.
Overwork, effect of, on heart, 30.

Pain, cause of, in gout, 168.
Palpation of heart, 52.
Palpitation, 63.
    prognosis of, 289.
Paraldehyde as a hypnotic, 284.
Percussion of heart, 52.
Plethora, 30.
Poisons, various, slow the heart, 105.
Precordial anxiety, 35.
    prognosis of, 287.
Precordial pains not always anginous, 116.
    many varieties of, 117.
Prognosis in heart affections, 287.
Puberty, cause of (note), 12.
Pulse and blood require attention, 224.
    during *tremor cordis*, 65.
    in tachycardia, 69.
    in Graves' diseases, 71.
    in palpitation, 63.
    normal, sometimes unusually slow, 96.

Raynaud's disease (note), 85.
Regurgitation, aortic, how brought about, 57.
    ventricular, Krehl's account of it, 60.
Remora of serous plasma in inter-vascular spaces, 27.
Rest, importance of, in treatment of senile heart, 234.
Rhyngrave, the, his cure for gout, 171.

Sclerosis, coronary, its relation to angina, 125.
Scott, Sir W., on *tremor cordis*, 66.
Senile degeneration of the heart from the morbid anatomist's point of view, 32.
Soda as antacid, 278.
Sound, booming first, what it signifies, 54.
    a booming second, what it signifies, 57.
Sounds of heart, progressive alteration of, as dilatation proceeds, 54.
Spa treatment, danger of, in senile heart, 251.
Spinal accessory, compression of, slows heart, 102.
Sphygmogram of hemi-systolic bradycardia, 106.
    of true bradycardia, 107.
    of feeble and irregular pulse, 46.
    of irregular pulses in dilated hearts, 47.
    of pulse of tachycardia, 78.
Stimulant, hot water the best cardiac, 244.
Stimulants, vascular, their use, 266.
    drugs that are, 275.
Strophanthus, use of, 267.

Strychnine as a heart tonic, 268.
Sympathetic, the katabolic nerve of the heart, 38.
Symptoms, objective, of senile heart most reliable, 224.
Syndrome of Graves' diseases, 71.
　of tachycardia, 71.
　of true angina, 124.
Systole shortened in tachycardia, 79.

Tachycardia, or heart hurry, 69.
　prognosis of, 289.
　treatment of, 292.
　physiological, 70.
　pathological, 72.
　from poisoning, 75.
　reflex, 82.
　action of augmentor nerve in, 83.
　often accompanies mitral stenosis, 74.
　sometimes emotional, 85.
　two cases of, 85.
Tea, tobacco, etc., as causes of angina, 127.
Temperance in all things important preservative of cardiac health, 236.
Temple, Sir W., on the treatment of gout, 170.
Tissues condense with age, 16.
Tithonus a typical aged man, 16.
　dies in real life, 17.
Thrombosis of veins common in gout, 167.

Thrombosis, source of many accidents in gout, 173.
Tobacco, use of, 254.
　dangerous in senile heart, 255.
Townsend, case of Colonel, 67.
Tremor cordis, 64.
　often arises from flatulence, 68.
　is never emotional, 68.
　prognosis in, 289.
　sudden onset of, 65.
　treatment of, 292.
Treatment of myocardiac failure, 231.
　of various cardiac symptoms, 292.
Turgescence, red, in gout, its cause, 168.

Uric acid decomposes copper of Fehling's test, 190.

Vagus, the anabolic nerve of the heart, 39.
　compression of, produces tachycardia, 90.
Vascular stimulants, action of, 275.
　must be combined with cardiac tonics in treating senile heart, 266.
Venosity of blood, cause of, 160.
Vichy water as an antacid, 278.

Water, hot, sipping, the best stimulant for the heart, 244.

www.ingramcontent.com/pod-product-compliance
Lightning Source LLC
Chambersburg PA
CBHW022106230426
43672CB00008B/1296